Water Resources Monograph 15

INFILTRATION THEORY FOR HYDROLOGIC APPLICATIONS

Roger E. Smith

with Keith R. J. Smettem, Philip Broadbridge and D. A. Woolhiser

American Geophysical Union
Washington, DC

Library of Congress Cataloging-in-Publication Data
Smith, Roger E., 1941-
Infiltration theory for hydrologic applications/Roger E. Smith; with Keith R. J. Smettem, Philip Broadbridge, and D. A. Woolhiser
p.cm.-- (Water resources monograph ; 15)
Includes bibliographical references
ISBN 0-87590-319-3
1. Groundwater flow. 2. Seepage. 3. Soil percolation I. Smettem, Keith R. J. II. Broadbridge, Philp, 1954- III. Title. IV. Series
GB1197.7.S6 2002
551.49--dc21 2002066604

ISBN 0-87590-319-3
ISSN 0170-9600

CONTENTS

Preface

Many recent books on soil physics provide a good coverage of unsaturated soil water flow from the perspective of water movement within soil. But those texts rarely deal with those parts of porous media hydraulics which inform a scientific approach to infiltration for hydrology, which is focused on the intake and movement of water at the boundary. Recent textbooks on hydrology, on the other hand, mostly continue to treat infiltration by typically reviewing the concepts of Green and Ampt, and listing one or more popular algebraic formulas (with parameters of limited physical significance) which are 50 or more years old. The concepts which link soil physics and hydrology and unify our understanding of infiltration from both rainfall and from irrigation conditions are usually not yet being presented as a unified body, but remain separately in the technical literature.

In this work we hope to fill what we view as a small open space in hydrologic literature: to provide a reference or guide for those interested in modern infiltration theory, to present the theoretical and mathematical basis of physically-based infiltration functions, and to indicate how to apply the theory to various hydrologic problems. It is assumed that the reader has an understanding of mathematics including appreciation of the basics of partial differential equations. However, mathematical complexities, especially lengthy derivations, will be avoided when not needed, especially where available in referenced material. The work outlined in Chapter 4 is important in support of the quality of the approximations used in Chapter 5, and for other reasons, but it is not necessary for the reader to understand all the mathematical complexities referred to there in order to understand the development of infiltration models in Chapters 5 and 6.

Other scientists are more qualified to write on the mathematics of soil physics; it is important to note that the effort here is hydrologic—to present the theory supporting a scientifically valid approach to hydrologic problems which involve infiltration. The contributing authors support this work with several chapters reflecting significant expertise and experience in both mathematical theory and field applications. But as in most scientific efforts, the work is built on the contributions of many earlier investigators, some of whose names are referenced herein. John Philip and Yves Parlange deserve prominent mention. It is on their work that much of the theory presented here depends.

Roger E. Smith

Acknowledgments

This manuscript was begun during a brief but fruitful visit to the Centre for Water Research, University of Western Australia, which was made possible by a Gledden visiting scholar Fellowship. Facilities were kindly provided by the Centre, and the cooperation and support of Prof. M. Sivapalan and Prof. Jorg Imberger as well as many other members of the Centre staff and students is gratefully acknowledged.

The contributions of contributing authors Phil Broadbridge, Keith Smettem, and David Woolhiser are important parts of this manuscript, without which it would be far less useful. The cooperation of Prof. C. Corradini at the Istituto di Idraulica, Facolta di Ingegneria, Universita de Perugia, and Florisa Melone at the IRPI, CNR, Perugia, Italy have been very helpful to the material presented in Chapter 7, and many discussions with them over the years have contributed significantly to the material presented here.

Finally, the facilities and staff of the Engineering Research Center, Colorado State University, have contributed to the preparation of the manuscript, for which I am indebted. Staff of the American Geophysical Union have also been helpful and patient in the aiding the preparation of the manuscript.

1

Introduction

A prominent feature in much hydrologic research and practice is a focus on groundwater and its quality. Given the time scale of groundwater movement and the current magnitude of industrial pollution, this focus may continue for years to come. More traditionally, hydrologists have often concerned themselves with the portion of rainfall that become streamflow: floods and their associated damage. At the same time, most rainfall over Earth's landmass becomes evapotranspiration. With only local exceptions, plants and crops are larger players in the hydrologic cycle than are rivers. Of course, rainwater at the soil surface either becomes soil water or surface runoff water, which can then reach the stream system. Soil water is the source of plant transpiration. From this perspective of the division of rainwater, infiltration processes are a very important part of the physics of the hydrologic cycle. This is no less true for that portion of the soil to which we add water ourselves when irrigating crops. Infiltration is thus a process of concern not only to hydrology but also to agriculture.

The infiltration of rain and surface water is controlled by many factors, including soil depth and geomorphology, as well as soil hydraulic properties and rainfall or climatic properties. Humanity has understood for ages that rainfall wets the soil and that rainfall may produce runoff. Our understanding of the physics of the process and the dynamics of porous media hydraulics has come rather recently. In fact, our ability to mathematically describe the response of a soil to a rainfall at its surface, and to understand the parameters that affect it, has arisen only in the last few decades.

BRIEF HISTORY

In 1911, Green and Ampt published a prescient paper describing a remarkably insightful picture of the process of infiltration from a ponded surface condition. Decades would pass before hydrologists attributed to this work the significance it deserves. In 1933, Robert A Horton, a consulting engineer and an active member of the small group forming the initial nucleus of the Hydrology Section of the American Geophysical Union [AGU], published his assessment of the role of infiltration in flood generation. At that time, engineering hydrologic

Infiltration Theory for Hydrologic Applications
Water Resources Monograph 15

practice treated the relation between rainfall and runoff with a very general "hydrograph separation" method, paying scant attention to process dynamics. Admittedly, the science of soil water flow was limited until after mid century when John Philip, Wilfred Gardner, and others made significant leaps in establishing soil physics. [As most readers will be aware, soil physics is a branch of soil science largely devoted to flow of water in soil.] Remarkably, however, even today engineering practices persist that are similar to those of Horton's time: approaches to hydrology that ignore the scientific knowledge of the past 60 years.

Horton first conceived "infiltration capacity" as a hyetograph separation rate that was generally applicable as a threshold for application to a rainfall intensity graph: the threshold intensity level being affected by soil conditions, seasons, and other phenomena [Horton, 1933]. A few years later, Horton refined this "capacity" concept by referring it to an infiltration rate that declines exponentially during a storm, and published a conceptual derivation of the exponential decay infiltration equation [Horton, 1936]. Although Horton seemed unaware of the work of Green and Ampt when he made his refinements, the difference between the two is noticeable: Horton was concerned with infiltration from rainfall while Green and Ampt treated infiltration from a flooded or "ponded" surface condition. More decades would pass before the relation between the soil water dynamics from these two boundary conditions was established and the potential unity of the mathematical description of the processes became clear.

The last half of the 20th century has seen great strides in soil physics, some of which have informed new approaches to the treatment of infiltration in hydrology. I will not attempt to summarize this recent scientific history here, lest an important brick in the structure go unmentioned. However, I cannot omit pointing out the significance of the mathematical work of John Philip and Yves Parlange, as mentioned briefly in the Preface. As will become transparent to the reader, the bulk of the material in the following chapters owes much to the insights of these two scientist/mathematicians. It is in part our attempt to formulate a hydrologically oriented and coherent presentation of their (and many others') contributions that we have written this monograph. Readers interested in a more mathematical and soil physics-oriented approach are encouraged to study, for example, the 1969 work of John Philip, as referenced. Very little of this material is out of date today. Also deserving mention is the recent breakthrough in the form of an analytic solution to a very realistic description of soil characteristics by Ian White and Philip Broadbridge, which is summarized in chapter 4. Parlange and others have also written extensively regarding this approach, which serves to substantiate the approximate analytic infiltration models presented below. Another recent advance heretofore unavailable are robust numerical solutions to the partial differential equations used to describe convective-diffusive flow, and the ubiquitous and ever-faster personal computer that has made such solutions practical. Numerical solutions, which appear in several places within the present text, are used here as demonstrations of the validity of mathematical assumptions used to produce analytic infiltration models.

RUNOFF MECHANISMS

In general, runoff at a particular location may occur from two types of soil hydraulic limits; in either case the soil surface will be saturated during periods of runoff generation. These two processes are now well understood by hydrologists. One is often, but not always, able to determine the dominant mechanism of a particular site and climate. In one case, for relatively lower rates of rainfall characterizing humid climates, the soil may saturate from below when downward unsaturated flow is limited by some restrictive subsoil or bedrock layer. Alternatively, the rainfall rate may exceed the rate at which the soil can accept input at the surface boundary - the soil becomes saturated just at the surface and runoff ensues. The first cause is hydrologically important in many areas, but involves factors of topography, soil depth, soil horizonation, as well as soil hydraulics. This mechanism is called saturation excess, and it is a case of *subsurface soil control.* The other runoff mechanism is *surface soil control*, and is the runoff process to which we apply Horton's name today. The present work addresses our modern understanding of this process, which owes much to the science and mathematics of soil physics.

At a much larger space and time scale, the interplay of rainfall rate and soil intake rate is illustrated schematically in Figure 1.1. For some climates and soil regimes, the dotted line may crudely represent the overall time (and/or space) probability distribution of rainfall rates, and the dashed line the geographical (and temporal) distribution of surface soil intake rates (specifically undefined here). While there is some overlap, the general dominant runoff mechanism is subsurface control: rainwater fills the soil profile above some limiting layer, and runoff follows from the exhaustion of soil storage.

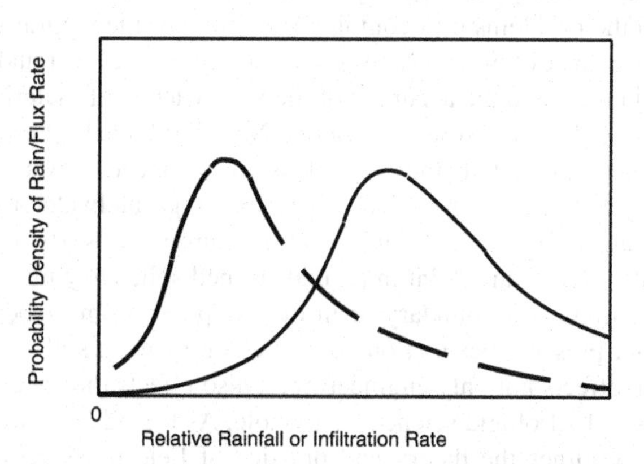

Figure 1.1. Rainfall rates and infiltration rates may each be thought of as having probability distribution. In humid climates with porous soils and low rainfall rates, the distribution of rainrates would be the curve on the left, and the infiltration rates distributed on the right. In other climates with fine textured soils and more intense storms, the two distributions would be reversed.

On the other hand, in areas (or seasons) characterized by surface control (the *Horton mechanism*) the two conceptual probability distributions would be interchanged: rainfall would more often come at rates in excess of the soil intake capacity. Here the control is at the surface, and comes from the interplay of the rate of supply and the limiting rate of intake. These illustrations are generalized and simplistic, and would involve greater complexity for an application to any region or climate. Nonetheless, the concept of the probabilistic interplay between soil hydraulic and rainfall intensity characteristics is useful when considering the spectrum of runoff mechanisms and conditions. Eagleson (1978) (and several after him) has dealt most completely with the probabilistic aspects of the Horton mechanism in this context.

ENGINEERING AND SCIENTIFIC APPROACHES TO HYDROLOGY

Philosophically, engineering is regarded as a technical application of science to human problems. In practice, certainly in the case of hydrology, the integration of scientific knowledge with engineering techniques is fraught with many limitations. The engineer must often apply her or his expertise with methodologies dictated by one or more regulatory agencies, must guard against liability claims, and should not be expected to have scientific currency in all areas. Within hydrologic science, other factors come into play, further restricting the application of science by engineers, including the burdensome cost of gathering data on soil and catchment properties. Complexity and uncertainty are additional factors involved in engineering problems that concern surface or subsurface water resources.

In this light, hydrology is arguably one of the branches of civil engineering most subject to the problems of a complex and unknowable natural system. For example, consider the differences between the structural engineer and the hydrologic engineer in regard to the accuracy of their knowledge of materials, internal state conditions, and system forces or inputs. No other branch of engineering is asked to design or predict in the face of as much uncertainty as hydrology. Furthermore, hydrologists, especially soil water or groundwater hydrologists, have no practical field methods by which to measure what is usually most crucial: the flux of water at any point in an undisturbed soil. They must depend on measuring conditions at a boundary or in the sampling of an associated value, such as the water pressure head, at one or more points in the soil. This dilemma in measurement effects not only engineering; it also effects the pace and character of advances in hydrologic science as a whole. As regards soil water infiltration, Chapter 8 outlines the theory and practice of field measurement, so that readers can gain some understanding of the relation between theory and measurement. This relationship is today made somewhat more fruitful via robust numerical solution methods now available to simulate the multidimensional flow conditions that field measurement usually create.

With the foregoing in mind, it is nevertheless true that modern hydrologic textbooks and recent hydrology handbooks do not well reflect the current understanding of infiltration theory. The methods on which practicing engineers depend are often those approved by regulatory agencies and are usually somewhat obsolete and empirical in basis—they do not reflect our current more sophisticated understanding of this part of the hydrologic cycle. Finally it must be acknowledged that this is in part due to the common paucity of appropriate rainfall data with which to apply infiltration theory.

OBJECTIVE AND SCOPE OF THIS WORK

This monograph presents the basis of our current understanding of soil infiltration theory while demonstrating its application to a variety of rainfall and soil conditions. It is not intended as an exhaustive treatise; rather it attempts to organize in one text major pieces of infiltration theory scattered through scientific journals—some of which are widely referenced, but most of which have yet to appear in textbooks.

Because infiltration theory grows out of soil physics and these two approaches are so closely related, Chapters 2 and 3 introduce porous media concepts, unsaturated flow, Richards' equation, and its mathematical features. Soil physics is typically focused on solving Richards' equation [see Chapter 3], which describes flow for a region within a porous media, while the infiltration theory presented here is based on an equation for flow continuity across the soil surface. Both assume the validity of Darcy's law. Whereas soil physics can be seen as more concerned with the effect of boundary conditions (and plants) on internal soil water dynamics, infiltration theory focuses on the effects of soil water dynamics on surface boundary conditions.

Our emphasis is on one-dimensional, vertical infiltration such as for the rainfall-on-soil condition. In explaining the theory and approximations, the mathematics of infiltration in the horizontal (gravity-free) [sorption] case is used to introduce concepts that can be extended to the vertical case. Reference is made to other infiltration cases, including two and three dimensional flow geometries, particularly in regard to field measurement methods. There are example applications at appropriate places in order to illustrate infiltration calculations in practical situations, but an extensive set of student problems are not included. The authors assume, tacitly, that applications of this material are not usually going to be in the form of hand calculations, but rather incorporation in hydrologic models. Thus extensive examples of such calculations are not provided. On the other hand, it is worth pointing out that while numerical solutions of Richards' equation are available for a greater variety of geometric conditions, the brute-force numerical approach to hydrologic problems is not universally robust and will not soon replace the use of simple analytic approximations in hydrologic modeling.

Therefore, the final product of this text is in the form of concise and robust equations (infiltration models) that can be applied in hydrologic models at a variety of scales for a variety of objectives.

2

Basic Porous Media Hydraulics

In this chapter, basic concepts of soil flow hydraulics are described in order to introduce concepts and definitions that are required to understand the material in later chapters. The description here is not intended to substitute for a more complete text on porous media hydraulics or soil physics. Good textbooks on those subjects are available, such as those of Jacob Bear [1972], A.T. Corey [1994], and Daniel Hillel [1980], among others. The flowing liquid in this discussion is assumed, unless otherwise indicated, to be water, but flow in porous media includes other fluids and also two-phase systems of a gas and a liquid

CAPILLARY PROPERTIES OF SOIL WATER

Capillary pressure occurs at the interface of two liquids within a porous media, and is a property of the liquids and their interface. Here in soil we are generally concerned with water and air. Water is considered a wetting fluid, because at the air-water-solid interface it is attracted more strongly to the solid medium. Air is the non-wetting fluid, with a pressure p_{nw}. There is a pressure discontinuity at the air-water interface that is balanced by a film tension at the water surface called capillary tension: $p_c = p_{nw} - p_w$. This creates a negative pressure in the water phase, p_w, compared with the air phase, and results in film curvature at the interface attachment to the solid boundary. The classical example of capillary interfacial force is the rise of water in a small capillary tube, in which the pressure difference at the capillary interface causes a rise of water in the tube of radius *R*. The force at this geometrically idealized interface is a function of the effective capillary radius and the interfacial tension, τ. As illustrated in Figure 2.1, with an assumed angle of wetting, *a*, the balance of forces can be used to show that the capillary rise, *h*, is

$$h = \frac{2\tau \cos\alpha}{R g (\rho_l - \rho_a)} \tag{2.1}$$

Infiltration Theory for Hydrologic Applications
Water Resources Monograph 15

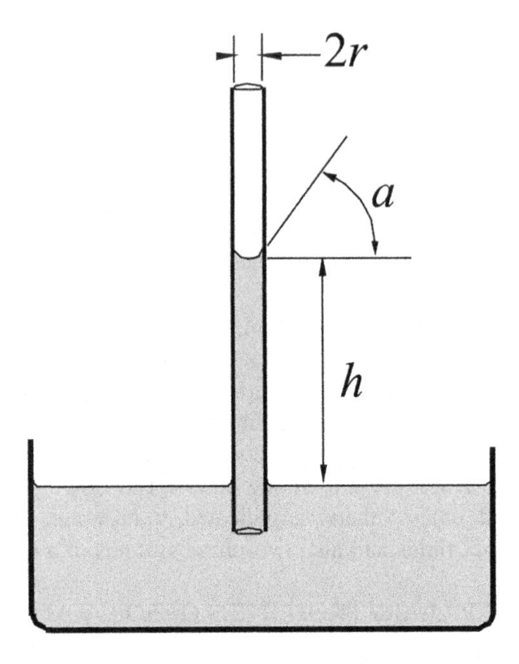

Figure 2.1. The rise of a liquid with capillarity in a small diameter tube is here illustrated to show the basic relations of capillary rise and pore radius.

where g is the gravitational constant, ρ_l is the liquid density, and ρ_a is the air density. Balance of static fluid forces require that the pressure head in the tube at the level of the water surface be zero, and that just below the interface the water in the tube has pressure head $-h$.

While Equation (2.1) is somewhat oversimplified, and various chemical factors affect the value of τ, the inverse relation of h to R is important. For water at 20 °C the relationship is, for h and R in cm., approximately

$$h = \frac{0.148}{R}$$

In a soil a wide variety of equivalent pore sizes exist, and the water occupies a limitless variety of shapes rather than tubes. Thus an enormous number of different interface configurations will act together to affect the distribution of water in an unsaturated soil. However, the finer the soil particle sizes, the smaller the effective values of R, and the greater the expression of water capillarity. The value of h is a height of water as calculated in this equation, and water behind the meniscus has a negative pressure with respect to the air outside. We can speak of capillary tension, or refer to pressures with negative values. Usually, soil water (negative) pressure is expressed in terms of equivalent water column height, as in Equation (2.1), but all the units shown in Table 1 are presently in use. One finds in the literature use of several terms: *capillary tension* (usually as a positive value), capillary *pressure potential,* soil water *potential,* or *head.* Units of cm are also used, but mm and m are SI units. Rather than h, we will more commonly refer to the soil water capillary *potential,* with the conventionally used symbol ψ, and it will commonly have a negative value.

TABLE 1 Conversions between common units for expressing soil water pressure or potential at standard condition.

1 of these units:	equals:	KPa	mm H_2O	Bar
KPa (kilopascals)		1	101.97	0.01
mm H_2O		0.009807	1	0.000098
Bar		100.0	10197.2	1

POROUS MEDIA

Porous media include any material, rigid or deformable, made up of small scale interconnected pores through which liquid may flow. By small scale, we mean that the pores are sufficiently numerous with respect to the scale of the flow region that the bulk flow through the material may be treated as a continuum fluid process. For unsaturated flow the pore sizes are generally assumed to be sufficiently small that capillary pressures are significant. A very coarse gravel may not exhibit all of the expected properties of capillarity at larger relative water contents, and thus may not resemble an unsaturated porous media such as soil. A single-sized capillary tube is not a satisfactory model for the capillary behavior of soils, and a better conceptual model would be a bundle of capillary tubes of a range of diameters reflecting the pore size distribution of the soil. This conceptual model is, however, unsatisfactory because a soil is composed of extremely irregular pore spaces between particles. Soil is more like a rigid sponge than bundle of tubes, with interconnections in all directions, and a variety of possible flow paths between two points in the medium. The soil can have a wide range of water contents and pressure potentials. For experimental purposes when microscopic uniformity is required, uniformly sized glass beads are often used, representing an extreme of uniformity in porous media. Real soils are rarely very uniform in particle size composition. The most uniformly sized soils are those where the sorting processes of wind or water deposition were involved in the soil creation or translocation.

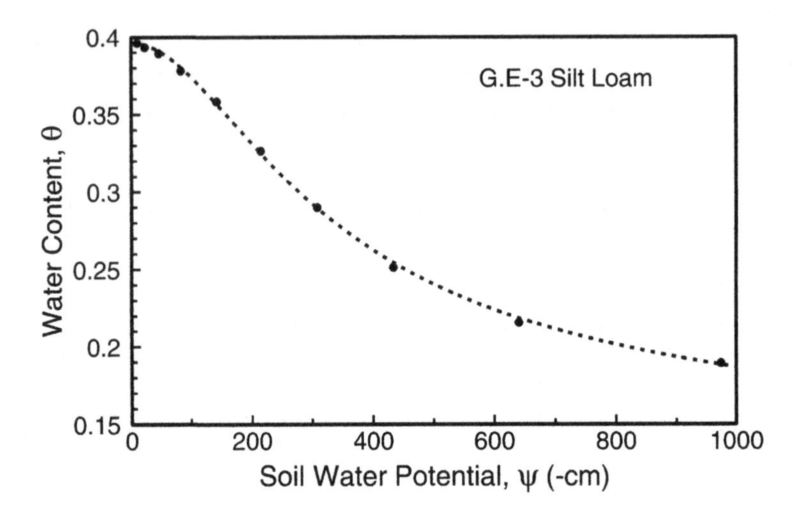

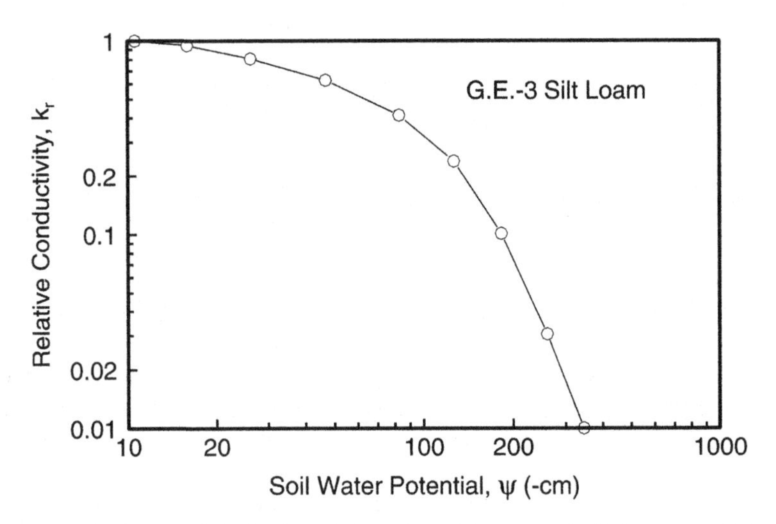

Figure 2.2. The basic soil hydraulic relations for a silt loam soil [Reisenauer,1963]. The water retention relation (a) plotted on an arithmetic scale, and (b) the hydraulic conductivity as a function of soil water potential plotted on a logarithmic scale.

The porosity of a porous media is the volumetric fraction of open pores (generally assumed interconnected). The theoretical upper limit of saturation is equal to the porosity. Each soil water content less than saturation has an associated value of (negative) capillary potential, which decreases (algebraically) with decreasing water content. When the capillary potential ψ is plotted against water content, θ, a relation such as Figure 2.2a is found, often called the *retention curve*. The S shaped curve depicted here is rather typical in general features, with water content decreasing least rapidly at very small and very large values of ψ.

As all soil physics textbooks will point out, there is often a different relation when water is being added to the soil (wetting), compared with the case where it is being withdrawn (drying). This complication to the θ-ψ relation is called soil water *hysteresis,* and is illustrated schematically in Figure 2.3. This results from pore geometries which require less energy for water to replace air than for water to be withdrawn. Since our purpose here is primarily to look at the intake of water at the soil surface, with usually monotonically increasing water contents, this hysteretic feature of soils will not be dwelt upon. Soil water hysteresis is potentially significant in cases of interrupted infiltration with significant redistribution of soil water.

Soil water contents are generally expressed in volumetric terms, and the value of θ is a volumetric fraction. The lower limit, θ_r, is never 0, except when the soil is dried in an oven, as there persists some water which cannot be removed from a soil by pressure gradients. Water can also migrate at very small saturations in the form of vapor, generally in response to temperature (i.e. vapor pressure) gradients. For practical cases of liquid water movement of interest here, this lower limit will be called θ_r, the *residual water content.*

The upper limit of θ for purposes of water flow and soil hydrology is likewise somewhat less than the porosity, and in this manuscript will be called θ_s, the *"saturated" water content.* It is necessarily less than the porosity, except where

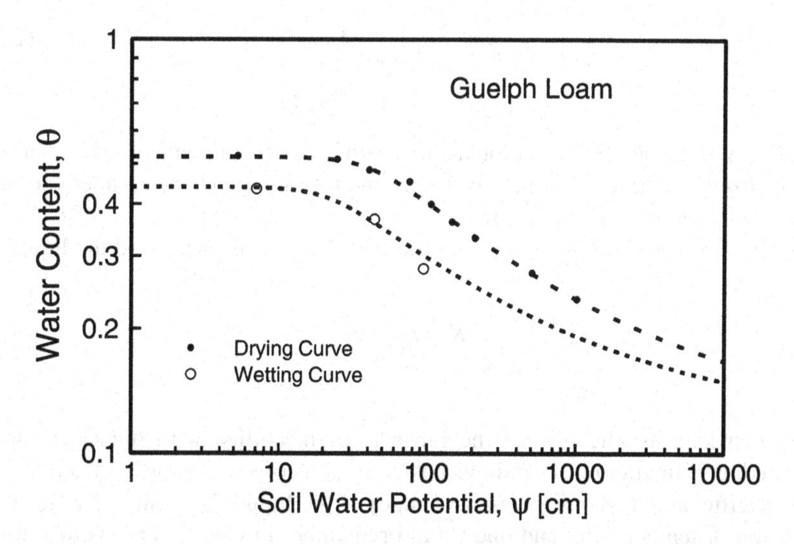

Figure 2.3. The retention curve for Guelph Silt Loam [Elrick and Bowman, 1964] demonstrates the hysteresis often observed in the relation between soil water potential and water content, which is a difference between the relationship when the water is drying and that when it is wetting.

imposed in the laboratory, because water entering a soil invariably traps some air in pores that are surrounded by water and the air does not escape. Thus all the pores of the soil are almost never truly "saturated," but we rather define θ_s as the water content at $\psi = 0$.

DARCY'S LAW

Flow in porous media, whether saturated or unsaturated, is assumed (and consistently shown) in almost all cases to be described by Darcy's Law. This basic relationship simply states that flux, or flow per unit area, *v*, is proportional to the gradient of the total potential, H:

$$v = -K \frac{dH}{dx} \tag{2.2}$$

in which K is the coefficient of proportionality, H is the total energy potential, comprised of the capillary pressure potential plus the gravitational potential. x is the measure of distance in the direction of flow. If we adopt the convention of measuring z positive downwards, (looking forward to infiltration calculations) then $H = \psi - z$:

$$v = -K\left(\frac{d\psi}{dz} - 1\right) \tag{2.3}$$

for Darcy's law applied to vertical flow with z positive downward. K is called the *hydraulic conductivity* of the medium, and for saturated flow is called the *saturated hydraulic conductivity*, K_s.

K expresses combined properties both of the porous media and the liquid:

$$K = k_p \frac{\gamma_w}{\mu} \tag{2.4}$$

in which k is usually termed the intrinsic permeability, with units of length squared, μ is the liquid dynamic viscosity, in units of force*time/area, and λ_w is the specific weight [force per unit volume] of the liquid. Viscosity of water is a function of temperature, and one should remember in the discussions to follow that temperature will affect conductivity and thus infiltration.

When the soil is not saturated, and for values of ψ less than 0., K in Equation (2.3) is the *unsaturated hydraulic conductivity*, and is a function of ψ. This introduces the basic and important relations between K, θ, and ψ, called the *hydraulic characteristics*, to which we now turn our attention.

HYDRAULIC CHARACTERISTICS OF POROUS MEDIA

As ψ becomes smaller (algebraically lower), the water content is reduced, while K is reduced very rapidly and by several orders of magnitude. This relation is shown in Figure 2.2b and 2.4, with a shape similar to the retention curve shown earlier. One reason for the reduction can be visualized by returning to the analogy of capillary tubes. High negative values of ψ represent flow in smaller and smaller 'tubes', which have a flow 'resistance' proportional to the 4th power of the radius, according to Poiselle's law. But also involved in this severe reduction of K is the increased tortuosity of the flow paths the water may take in flowing through the complex shapes made up by the soil particle interfaces.

Alternatively, since there is a relation between ψ and θ, one can speak of the relation K(θ). This description of K is the second hydraulic characteristic of a porous medium, in addition to the retention relation: along with the retention relationship, either the K(θ) or the K(ψ) relation may be specified to define the unsaturated behavior of a soil. Figure 2.4 illustrates a typical K(θ) relation, whose shape is somewhat different from that of the other relations.

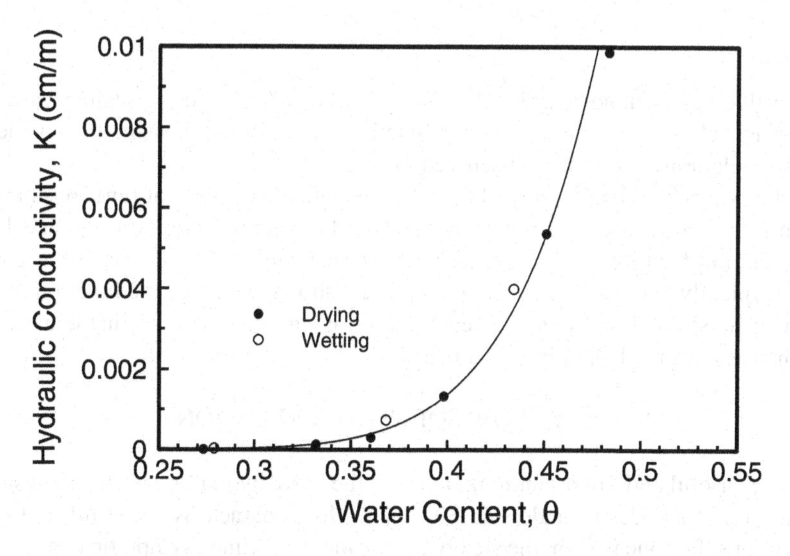

Figure 2.4. The relation between hydraulic conductivity and soil water content does not have significant hysteresis even if the retention relation is hysteretic, as illustrated here for Guelph Loam [Elrick and Bowman, 1964] (compare with Figure 2.3).

The relations between ψ, θ, and K are called the soil *hydraulic characteristics*, and are fundamental descriptors for the porous media flow properties of the soil. Figures 2.2 through 2.4 illustrated typical relationships between these characteristics. The functional or mathematical descriptions of the hydraulic characteristics, discussed below, are called *characteristic relations*. In each case, the relationships are monotonic: K and θ always decrease (often rapidly) with (algebraically) decreasing values of ψ. The range of values for K and ψ usually covers many orders of magnitude, while the range of values of θ is generally much smaller.

Scaled Hydraulic Characteristics

The porous media hydraulic relations are more concisely described in scaled form. Water content is generally scaled between its limits θ_r and θ_s,

$$\Theta_e = \frac{\theta - \theta_r}{\theta_s - \theta_r} \tag{2.5}$$

while K is commonly scaled by the saturated conductivity K_s :

$$k_r = \frac{K}{K_s} \tag{2.6}$$

The value k_r is termed *relative hydraulic conductivity*. There are scaling values for ψ as well, but they depend on the function chosen for expressing the characteristic relations, and will be discussed later.

For various reasons, many of them mathematical, a variety of functions have been proposed or used for representing the relationships among Θ_e, ψ, and k_r. The relation for $k_r(\psi)$ is often called the *relative conductivity curve*. Both relations typically exhibit the reversal of curvature shown here, and the functions are somewhat simpler when presented on log-log graphs. Various mathematical forms are presented and discussed below.

STEADY FLOW SOIL WATER RELATIONS

It is useful and important to understand the basic hydraulics of steady unsaturated flow in order to understand unsteady flow characteristics of infiltration conditions. For one reason, the steady case is the large-time asymptotic condition for most infiltration boundary conditions found at the soil surface.

Static Soil Water Conditions If a uniform, isothermal soil profile of large depth has a water table at its lower end, it will, after sufficient time (ignoring evaporation), exhibit relative wetness which is inversely proportional to the height above the water level. The depth z is here measured positive downwards, and we will here define $z=0$ at the water table. The total energy potential, H, at any level, z, above the water level is ψ - z, where ψ is the (negative) soil water pressure potential. If there is no flow, then there is no gradient from Equation (2.3), and H is the same at all levels, and, so $\psi = z$. Thus the vertical change in water content, or the capillary rise profile will reflect the soil retention characteristic: $\theta(z) = \theta(\psi)$. Static conditions such as this may in some simple cases be used to determine the relation for $\theta(\psi)$.

Gravity-free Flow Darcy's Law for horizontal flow, unaffected by gravitational potential, is from Equation (2.2) simply

$$v = -K(\psi)\frac{d\psi}{dx} \tag{2.7}$$

A useful variable has been developed in connection with flow as described by this equation. This is the *flux potential*, which will herein be designated by the variable ϕ, and is defined as

$$\phi(\psi) = \int_{-\infty}^{\psi} K(h)dh \tag{2.8}$$

As evident from this expression, ϕ is dependent on the characteristic relation $K(\psi)$ that describes the medium. Its significance is that Darcy flow is described by the gradient of ϕ, thus:

$$v = -\frac{d\phi}{dx} \tag{2.9}$$

The flow or potential flow in an integral sense between two points in a soil characterized by the relation $K(\psi)$ can be expressed

$$v(a,b) = \frac{\phi_b - \phi_a}{x_a - x_b} = \frac{1}{x_a - x_b}\int_{\psi_a}^{\psi_b} K(\psi)d\psi = \frac{\phi_{b,a}}{x_a - x_b} \tag{2.10}$$

in which the definition for $\phi_{b,a}$ is implicit. Hereinafter this variable, when so subscripted, will be referred to as the *differential flux potential*. The flux potential concept is most often used in multidimensional flow expressions [Philip, 1969]. We will have occasion to use the relative flux potential later in calculating redistribution.

Flow in the Presence of Gravity In the soil column of the previous static example, there can also be either steady upward or downward flow, resulting in a depth and water content relation θ(z) which can be calculated from Equation (2.3) if the soil hydraulic relations are known.

With z negative upwards, downward flow is positive. (Any convention for defining z may be assumed without affecting the outcome, but one must be consistent.) The limiting value of downward steady unsaturated flow is $-K_s$, which is -K at $\psi = 0$, since otherwise ψ would have to increase into the positive range as z increased above the water table. For $|v| < K_s$, Equation (2.3) is rearranged to obtain

$$dz = \frac{-K(\psi)d\psi}{v - K(\psi)} \tag{2.11}$$

which may be integrated and expressed as

$$\int_0^{-z} dz = -\int_0^{\psi} \frac{K(h)dh}{v - K(h)} \tag{2.12}$$

in which *h* is the integration variable. Note that this relation reduces to the static case as *v* becomes 0. Also, as z becomes large, the relation correctly indicates that the value of K approaches *v*. This is the steady unsaturated flow condition, unaffected by a boundary: ψ becomes uniform at the value ψ(*v*), which is the inversion of the hydraulic characteristic relationship K(ψ) at K = *v*.

Given a relation for K(ψ), Equation (2.12) can in principal be solved for the profile ψ(z). However, most realistic forms of K(ψ) do not lend themselves to analytic integration of Equation (2.12). In the practical case it is simpler to solve the equation numerically in steps using Equation (2.11). This can be done using a simple spreadsheet program, for example. In this method, ψ is first stepped in small increments. An interval (arithmetic) mean value of ψ is then used to first find a mean effective K(ψ), and then to solve Equation (2.11) for value of the right side. This determines the Δz value, and thus the associated step in z. The procedure is repeated until z covers the range of soil height desired. One may start with ψ = 0 and z = 0 for i = 1, and the recursion formula is simply:

$$\bar{y} = 0.5^*(y_i + y_{i-1})$$

$$z_i = z_{i-1} - \frac{(y_i - y_{i-1})K(\bar{y})}{v - K(\bar{y})} \tag{2.13}$$

Note that ψ becomes more negative as z decreases (negative upwards), since the denominator of the last term is negative for all $v < K$, and z will decrease (upward) as ψ is reduced.

1. *Downward Steady Flow.* This is potentially important in infiltration problems, as a long term asymptotic condition, and we will revisit this in a later chapter as a problem for layered soils. Flow v is positive, but must be less than or equal to K_s. Thus, ψ will start at the water table as 0, and decrease only until $K(\psi)$ approaches v. As the denominator in Equation (2.13) approaches 0, the value of z will decrease more rapidly with each small decrease in ψ. The limiting condition is the steady value for unsaturated flow: a value of ψ such that $v = K(\psi)$. Figure 2.5 shows this asymptotic behavior for two values of v.

2. *Upward Steady Flow.* The case of upward steady flow is also not trivial in hydrology- it represents the case of water flowing upward in response to either

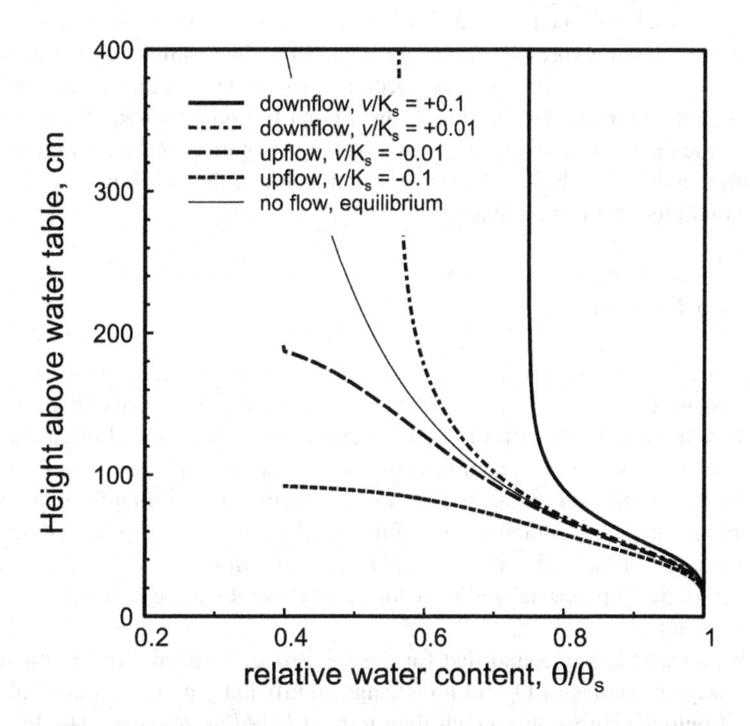

Figure 2.5. Steady flow upwards from or downwards to a water table exhibits characteristic water content relations. Upward flow has a limiting value dependent on the water table depth.

root water use or soil surface evaporation. In this case v is negative, and now there is no lower limit on ψ in recursion Equation (2.13) as K becomes smaller, since the denominator cannot approach 0. However, as K becomes increasingly small, the increase of z now becomes very small as the numerator becomes small, and there is a height $-z_L$ (depending on v) beyond which water may not be moved upwards because increasingly small conductivity cannot be compensated by higher gradients of capillary potential. One practical implication of the relations illustrated here, and shown graphically in Figure 2.5, is that for a given soil there is an interaction between the depth from the bottom of the roots to a water table and the maximum rate at which the roots can take water. Moreover, there is an upper limit on this upflow rate for any water table distance.

FUNCTIONAL REPRESENTATION OF HYDRAULIC CHARACTERISTICS

There have generally been two approaches to the functional expressions for porous media hydraulic relations, producing two kinds of approximations. The first kind, which has been perhaps less common in soil physics, is the attempt to find a function that expresses as closely as possible the results of experiments in which hydraulic characteristics have been measured. The second kind, which is often seen in the research literature, is an attempt to find functional forms which allow some mathematical advantage in attacking a solution of the flow equations presented in the next chapter. Here we present some of the first kind, and discuss the second kind in the next chapter.

Soil Water Retention

Figure 2.3, above, illustrates how soil water content is reduced as soil capillary pressure potential is reduced. One of the earliest mathematical expressions for this relation is that of Brooks and Corey [1964], which is still often used. It was developed more in connection with flow in porous sandstone than for soils, and its form is indeed more suited for very uniform material than for one with a wide range of pore geometries. For functional purposes, the water content is expressed in normalized form, the *effective saturation*, Θ_e, defined above in Equation (2.5). This normalized value of water content will be used often in subsequent chapters.

Brooks and Corey found that for the sandstone, sands and other soils they were studying, values of Θ_e did not change significantly until the potential was reduced below a threshold, which they termed *bubbling pressure,* ψ_B. Further, the relation of Θ_e to potential for values of ψ less than ψ_B they found to be well described by a power function of the value of ψ scaled by this value:

$$\Theta_e = \left(\frac{\psi}{\psi_B}\right)^{\lambda} \tag{2.14}$$

in which λ is termed the *pore size distribution* parameter.

A curvilinear form for $\psi(\Theta_e)$ more suited to soil measurements was presented by van Genuchten [1980]. The retention relationship of van Genuchten is:

$$\Theta_e = \left[1 + (\alpha_g \psi)^n\right]^{-m} \tag{2.15}$$

van Genuchten pointed out that the same relation with m = 1 had been used by many others earlier [e.g. Ahuja and Swartzendruber, 1972]. The parameter α_g is conceptually the inverse of ψ_B and has dimensions of L^{-1}. As an alternative or a simplification, van Genuchten suggested that m could be related to n, using m = 1 - 1/n. This has become a very popular simplification in the years since.

There have been many other suggested forms for the retention relation, but these will suffice for our use in the discussions to follow, and these can describe the soil retention in the vast majority of soils.

Unsaturated Hydraulic Conductivity

To obtain a relation for $K(\theta)$ or $K(\psi)$, Brooks and Corey used the ideas of Burdine [1953], which when applied to their assumptions produced a similar relation for *relative conductivity*, k_r,, similar to the retention curve:

$$k_r \equiv \frac{K}{K_s} = \left(\frac{\psi}{\psi_B}\right)^{\eta} \tag{2.16}$$

in which the conductivity exponent η was related to λ: $\eta = 2 + 3\lambda$. These two relations may be combined to describe the relation between relative conductivity and scaled water content:

$$k_r = \Theta_e^{\varepsilon} \tag{2.17}$$

in which $\varepsilon = \eta/\lambda = (2 + 3\lambda)/\lambda$. Figures 2.6 and 2.7 illustrate the fit of this relation to measured data from various soils. It is clear that the sudden start of the decay of θ at the value of ψ_B is an approximation that is better in some cases than

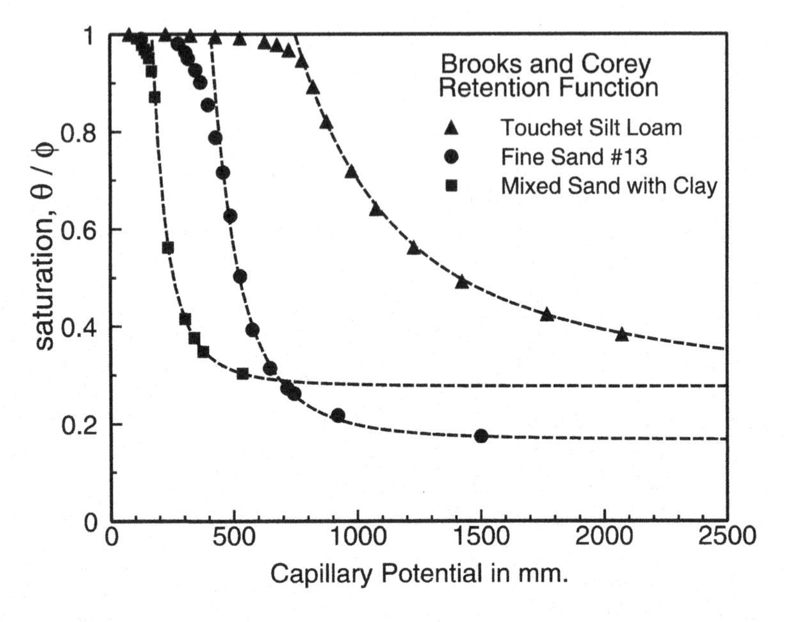

Figure 2.6. The Brooks and Corey retention function fitted to experimental data for 3 types of soils. The only part of the curve not well matched is the shoulder where the curve approaches saturation.

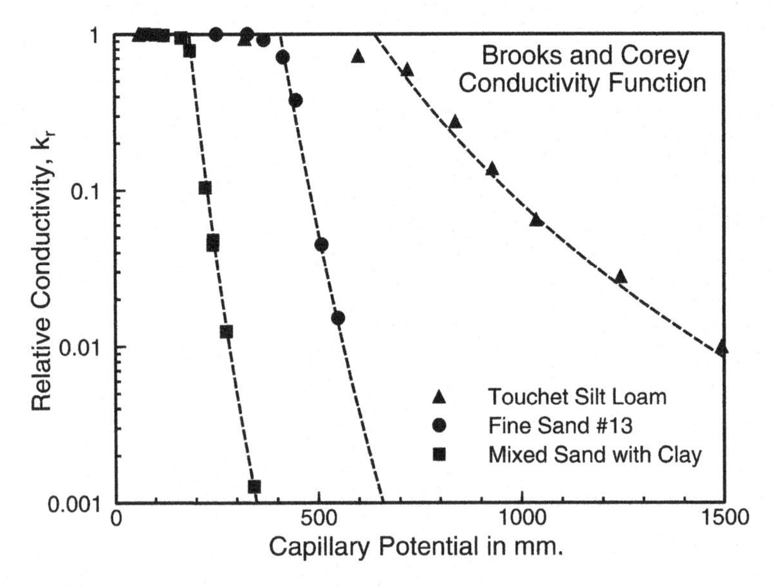

Figure 2.7. The relation of relative conductivity to capillary potential, Equation (2.16) from Brooks and Corey, fitted to the same soils shown in Figure 2.6.

in others. Another disadvantage of this relation is the difficulty that is introduced in solution of the unsaturated flow equation by the discontinuity in the slope $d\theta/d\psi$ at ψ_B. This is not necessarily a problem for conditions which do not reach saturation, but is a serious difficulty in hydrologic conditions of infiltration to ponding and runoff.

Van Genuchten adopted a functional relation between assumed soil pore geometry and the conductivity relation which produces a much more complex function for the conductivity relation, based on work of Mualem [1976]:

$$k_r(\Theta_e) = \Theta_e^{1/2}\left[1-\left(1-\Theta_e^{1/m}\right)^m\right]^2 \tag{2.18}$$

which can be used with Equation (2.17) to obtain the conductivity curve, $k_r(\psi)$:

$$k_r(\psi) = \frac{\left\{1-(\alpha_g\psi)^{n-1}\left[1+(\alpha_g\psi)^n\right]^{-m}\right\}^2}{\left[1+(\alpha_g\psi)^n\right]^{m/2}} \tag{2.19}$$

It is also possible to derive a somewhat simpler relation for $k_r(\psi)$ by using the Burdine pore conductivity theory rather than that of Mualem [van Genuchten, 1980]. Equation (2.19) is differentiable, which is important for numerical solution of transient flow equations, but is not easily integrable, which is a significant disadvantage for infiltration theory, as discussed below.

A Generalized Formulation

A function of the same mathematical form as that of van Genuchten, but retaining the physically meaningful parameters of Brooks and Corey in Equation (2.14) will be used often in this work. This may be referred to as the *transitional Brooks and Corey* (TB-C) relationship [Smith, 1990; Smith *et al.*,1993]:

$$\Theta_e = \left[1+\left(\frac{\psi+\psi_a}{\psi_B}\right)^c\right]^{-\frac{\lambda}{c}} \tag{2.20}$$

The parameters λ and ψ_B are the same as for the Brooks and Corey relations, and the parameter c has a defined effect on the shape of the curve near $\psi = \psi_B$ as illustrated in Figure 2.8. Note that this modification treats the part of the curves not well fitted by Equations (2.14) and (2.16) as shown in Figures 2.6 and 2.7. Parameter ψ_a is a small shift parameter, one use of which is the approximation of hysteresis, if desired, and may be left at 0. With $\psi_a = 0$, Equation (2.20) approaches the Brooks-Corey relation asymptotically as c becomes large. Likewise, the Brooks and Corey $k_r(\Theta_e)$ relation (2.16) is retained, and when combined with Equation (2.20), Equation (2.16) is altered to become

$$k_r(\psi) = \left[1 + \left(\frac{\psi + \psi_a}{\psi_B}\right)^c\right]^{-\frac{\eta}{c}} \tag{2.21}$$

Clearly, by comparison of Equation (2.20) to the van Genuchten relation Equation (2.15), $\alpha_g = 1/\psi_B$, $n = c$, and $m = \lambda/c$. Using the common simplification $m = 1 - 1/n$ in the van Genuchten relation will, however, contradict these 1:1 parameter interrelations. One of the features of Equations (2.20) and (2.21) is, in addition to their versatility, the fact that each of the parameters has a physically or functionally related meaning, and the parameters have relatively independent effects on the shape of the curve.

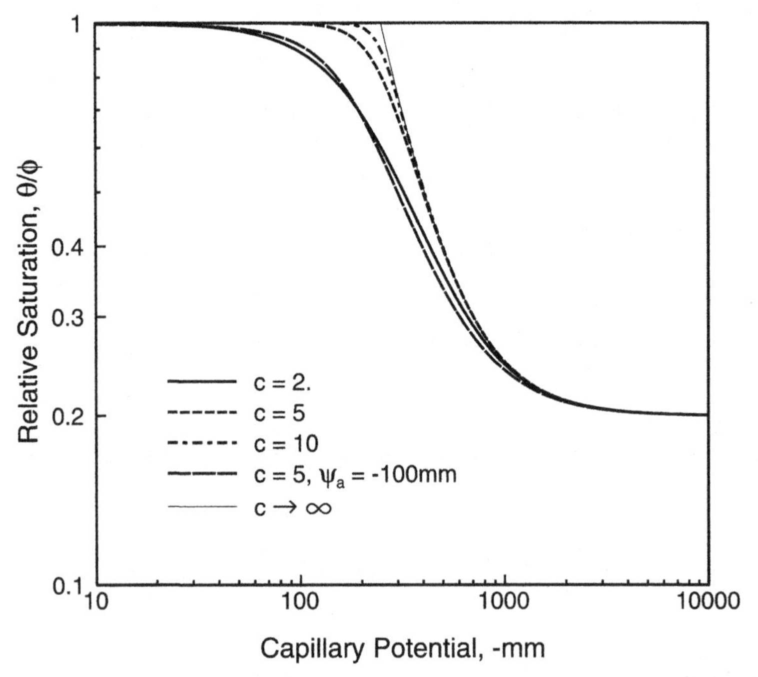

Figure 2.8. Illustration of the role of the curvature parameter, c, in the transitional Brooks-Corey soil characteristic functions. As $c \rightarrow \infty$ the function asymptotically approaches the original Brooks-Corey relations.

Other mathematical assumptions have been made regarding the soil hydraulic characteristics, usually for mathematical reasons for solution of the equation(s) for soil water flow, but which generally are not such good models of measured soil properties. Some of these are presented in the next chapters. In some cases, numerical solution of the flow equation can be done with use of tabulated values for the relations of θ, ψ, and k_r, finding values needed for computations by use of table look-up and interpolation. This will not allow us to explore the production of a hydrologically meaningful infiltration model, however. Indeed we will find that analytically derived infiltration functions are not particularly sensitive to small changes in the soil characteristics, but rather to integral properties of the hydraulic characteristics.

SUMMARY

Some of the basic principles of porous media hydraulics have been introduced here, including the basic hydraulic properties of porous flow media, and the basic equation for flow, Darcy's law, and its application to steady upward and downward flow. Mathematical expressions of the hydraulic properties of unsaturated soils have been introduced that are used in subsequent infiltration studies. In the next chapter, we will employ these concepts and hydraulic relations in the case of unsteady flow equations.

3

Some Essentials of One-Dimensional Porous Media Dynamics

INTRODUCTION

Many of the equations and concepts of classical one-dimensional porous media flow dynamics will be introduced in this chapter, because they are a part of the tools used in the following chapters when the focus is more directly on infiltration. Much of what is presented in the first part of the chapter may be found with greater detail in most soil physics textbooks. In addition, it should be understood that infiltration flux at the surface can be calculated indirectly by solving equations of porous media dynamics (usually numerically), although this does not lead directly to an infiltration model. Such numerical solutions will be used throughout this presentation to demonstrate various results and compare to approximations.

In the discussion below, cases of both infiltration and gravity-free absorption will be presented. The value of this is twofold, as should be appreciated later when analytic or approximate solutions are found. First, the adsorption case lends itself to analytic solution much more readily than the infiltration case. Secondly, the adsorption case is a valid approximation for infiltration in the earliest time period when adsorptive flux is much larger than the convective or gravitional flux. Absorption will be discussed first, followed by treatment of infiltration.

Definition

In this monograph the terms *imbibition* or *absorption* will be used to refer to the intake of water at a soil surface unaffected by gravity, such as horizontal flow, and *infiltration* will imply vertical downward flow including the gravitational gradient in Darcy's law. These two processes are very closely related, since the early stages of infiltration into a relatively dry soil are dominated by capillary potential gradients to such an extent that the absorption condition effectively describes the process during early times. The symbol v will be used to represent

Infiltration Theory for Hydrologic Applications
Water Resources Monograph 15

flux of soil water, volumetric flow per unit area. The mean or gross soil water *velocity* is the flux divided by the porosity; *i.e.* gross velocity = v/Φ, whereas the net velocity is the flux divided by the water area, v/θ.

The infiltration or absorption flux *at the intake boundary* (the soil surface) will be represented by the symbol f. By definition, with the subscript o referring to surface conditions, we have $f \equiv v_o$. For infiltration, the limiting value of f — the maximum vertical flux which the soil can accept at any time given an unlimited supply at atmospheric pressure — is termed the *infiltrability*, and represented by the symbol f_c. The corresponding flux limit for absorption will be termed *absorptibility,* to distinguish the two different flow equations. The value of f_c is a function of soil conditions and changes with time, as discussed below. The value of f, on the other hand, may be fixed if it is controlled by the rate at which water is supplied. Under rainfall of rate r [$r > K_s$], f will be fixed by r early in the event, and become equal to f_c later. It is the description of these processes that is the subject of this monograph.

Assumptions

The mathematical descriptions of soil water dynamics presented below make a set of simplifying assumptions. Later on many of these will be relaxed, or will be discussed and their validity with respect to natural soils will be demonstrated. The major assumptions made here allow us to introduce some basic mathematical descriptions that are important in understanding how infiltration functions are developed. Perhaps the most important assumption is that the movement of water into relatively dry soil can be described without explicit treatment of the flow of air. Infiltration or absorption involves the replacement by water of air in the soil pores. In laboratory conditions, one can provide an open or mesh bottom on a cylinder of soil to allow air an escape path. This is not necessarily the case when rainfall moves into the surface of a soil. Air can compress and cause a reduction in the pressure gradient across a wetting "front" of water entering a soil, and it also has a viscous resistance to movement through the soil ahead of entering water. However, the hydraulic potential gradient required to move air against the resistance of the soil pore structure is often quite small compared to the capillary pressure gradient, and there may (or may not) be a large reservoir of air to compress - *i.e.*, the depth of water entering may be very small compared with the volume of air. Nevertheless, air compression and counterflow, in some cases, will modify to some extent the results shown here. There are several studies on true two-phase flow to which the reader may refer, including McWhorter [1971], Brutskern and Morel-Seytoux [1970], Morel-Seytoux [1973], etc.

Soil is assumed homogeneous for the present chapter, but layered soils will be treated below. Initial soil water conditions are also assumed homogeneous, and the sensitivity of computations to this assumption will be demonstrated later.

The vapor-based movement due to thermal gradients is also ignored; the time scale for this process is much larger than for rain infiltration. The effects of soil swelling are also not treated here. There are several mathematical treatments of these processes in the literature. Another typical soil complication not treated here is soil water hysteresis, mentioned above, in which the relation of θ to ψ is different for water intake than for drainage. Insofar as simple infiltration conditions create soil wetting at all places, hysteresis would not become a factor without significant cycles of wetting and drying. For the present we deal only with wetting, and redistribution of water is discussed in Chapter 7.

ABSORPTION: GRAVITY-FREE INTAKE

To express the dynamics of soil water flow, we employ a fundamental dynamic continuity expression:

$$\frac{\partial \theta}{\partial t} + \frac{\partial v}{\partial x} = j \tag{3.1}$$

in which v is the Darcy flux, x is the directional distance, and t is time. The variable j is an external gain/loss rate such as root water use, which will be left at 0 for the present purposes. This equation merely expressed the idea that a change in storage at some location must be accompanied by a divergence (increase or decrease) of flow through the same point.

In the Darcy flux relation Equation (2.2), with the gravitational gradient removed, we have $v = -K\, d\psi/dx$. This flux expression, combined with Equation (3.1) for $j = 0$ becomes

$$\frac{\partial \theta}{\partial t} = \frac{\partial}{\partial x}\left(K \frac{\partial \psi}{\partial x}\right) \tag{3.2}$$

In order to form an expression in the form of a diffusion equation, for which a considerable history of mathematics is available, it is common to define a term that is known as soil water *diffusivity*, D:

$$D(\theta) \equiv K \frac{d\psi}{d\theta} \tag{3.3}$$

and to substitute this term into Equation (3.2) to form a θ-based form of the equation:

$$\frac{\partial \theta}{\partial t} = \frac{\partial}{\partial x}\left(D \frac{\partial \theta}{\partial x}\right) \tag{3.4}$$

Note that K in Equation (3.3) can be viewed as a function of either θ or ψ, but that D is treated as a function of θ. Transformation to the θ-based form removes the formal dependence on two variables (both ψ and θ). Since the form of Equation (3.4) is that of a diffusion equation, if D were constant in soils (as it is not), it would be amenable to treatment with the considerable mathematics of linear diffusion equations [e.g. Carslaw and Jaeger, 1959].

For all cases we assume for simplicity that the initial conditions are uniform:

$$t = 0,\ x > 0;\ \ \theta = \theta_i \tag{3.5}$$

where subscript i indicates initial condition (assumed uniform). For absorption from a step change in pressure potential at the soil boundary, the boundary condition is

$$t \geq 0, \quad x = 0; \quad \psi = \psi_o \tag{3.6a}$$

for the ψ-form equation (3.2), or

$$t \geq 0, \quad x = 0; \quad \theta = \theta_o \tag{3.6b}$$

for Equation (3.4).

An alternative important absorption boundary condition is the constant flux condition:

$$t \geq 0, x = 0, \quad v_o = -K\frac{d\psi}{dx} \tag{3.7a}$$

or in the θ-based form using Equation (3.3),

$$t \geq 0,\ x = 0, \quad v_o = -D\frac{d\theta}{dx} \tag{3.7b}$$

which specifies a fixed value v_o for the intake rate at x = 0.

Absorption with a Constant Head Boundary

As long as the form of $D(\theta)$ is well-behaved, Equation (3.4) can be reduced to an ordinary differential equation by the Bolzmann transformation, which presumes the solution $\theta(x,t)$ will scale as $t^{1/2}$, and therefore substituting the similarity variable $\varphi = xt^{-1/2}$ to obtain a transformed equation:

$$-\frac{\varphi}{2}\frac{d\theta}{d\varphi}=\frac{d}{d\varphi}\left(D\frac{d\theta}{d\varphi}\right) \tag{3.8}$$

with conditions $\varphi = 0$ at $\theta = \theta_s = \theta_o$ and $\varphi \rightarrow \infty$ at $\theta = \theta_i$. With the indicated transformation, the solution to Equation (3.4) subject to condition (3.5) and (3.6) is [Philip, 1957a, 1969]:

$$x(\theta,t)=\varphi(\theta)t^{1/2} \tag{3.9}$$

The value of $\varphi(\theta)$ is given by the solution to Equation (3.8), which depends on the soil hydraulic characteristics. Conceptually, one may think of the solution to Equation (3.4) as a wetting profile scaled on $t^{-1/2}$, as illustrated in Figure 3.1. Equation (3.9) falls short of being a true "solution," given the complexity of solving Equation (3.8) for φ. Philip[1969] solves Equation (3.8) by rewriting and integrating it to produce

$$\int_{\theta_i}^{\theta}\varphi\, d\theta=-2D\frac{d\theta}{d\varphi} \tag{3.10}$$

which he solves numerically.

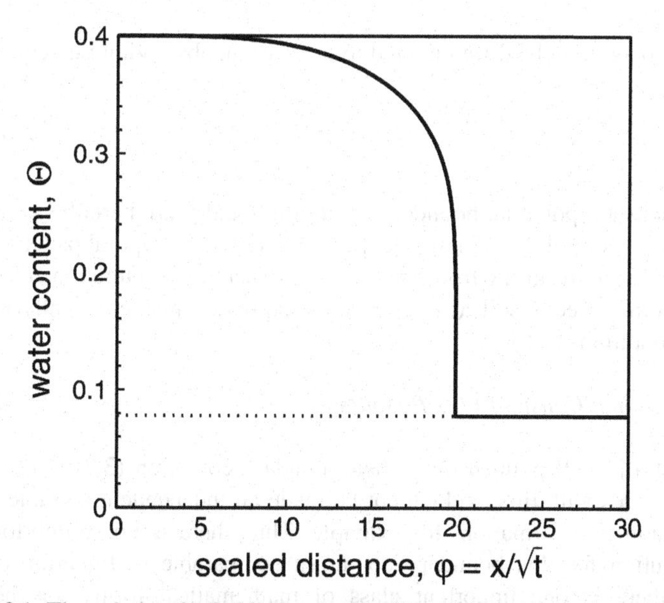

Figure 3.1. The solution for imbibition in $\theta(\varphi)$ is a single curve as a function of the Boltzmann transformed variable, and the advance of θ into the soil scales as the square root of time.

In regard to absorption at the boundary, the adsorbed amount, or depth, I (units of length), for Eq. (3.9) under the conditions (3.5) and (3.6) can be expressed directly as

$$I = \int_{\theta_i}^{\theta_s} x \, d\theta = t^{1/2} \int_{\theta_i}^{\theta_s} \varphi d\theta \tag{3.11}$$

The left integral is simply the area under the advancing "wave" of water, which must be the total infiltrated amount. The integral coefficient on the right side of Equation (3.11) is an important quantity in absorption and infiltration, called *sorptivity*, $S(\theta_i)$:

$$S(\theta_i) = \int_{\theta_i}^{\theta_s} \varphi d\theta \tag{3.12}$$

Sorptivity has the dimensions [$LT^{-1/2}$]. Graphically, it is the area under the scaled profile of $\theta(\varphi)$ between θ and θ_i, shown in Figure 3.1.

Equation (3.11) is an analytic equation, with no assumptions on the soil characteristics, for the absorption "depth" as a function of time:

$$I = St^{1/2}, \tag{3.13}$$

which, moreover, may be differentiated to produce an absorption rate equation:

$$f_c = \frac{S}{2} t^{-1/2} \tag{3.14}$$

Given the constant potential boundary condition, f and f_c are here the same, i.e. this describes absorptibility. These equations describe the temporal pattern of the absorption of water resulting from the sudden application of saturation at the surface of a homogeneous soil at time = 0, as expressed mathematically by the boundary conditions (3.6).

Absorption from a Constant Flux Boundary

As Philip [1969] pointed out, when boundary condition (3.7) is used, the range of θ varies with time and cannot be made an independent variable, as in the Boltzmann transformation, for example. Thus there is no straightforward general solution for this boundary condition comparable to Equation (3.11). However, there is one important class of mathematical forms for the soil hydraulic characteristics which allows transformation of nonlinear diffusion equations to a solvable form, which is discussed in detail in the next chapter.

Other approaches will be useful in regard to specifically evaluating the change of θ at the boundary with the infiltration integral in Chapter 5. Philip [1969] has identified a class of solutions, somewhat unwieldy, for the constant flux boundary that depend on having the diffusivity a power function of scaled θ.

INFILTRATION: VERTICAL FLOW

Combining Equation (3.1) with Darcy's law, Equation (2.2), and now including gravity, the equation for vertical, one-dimensional unsaturated flow is

$$\frac{\partial \theta}{\partial t} = \frac{\partial}{\partial z}\left(K \frac{\partial \psi}{\partial z} - K \right) \tag{3.15}$$

Here z is measured positive downwards from the surface. This is commonly known as Richards' equation, after L.A. Richards [1931]. Philip [1969] noted that, categorically, Equation (3.15) is a Fokker-Plank equation, and also pointed out the similar contribution of Buckingham [1907].

The general initial condition is as shown in Equation (3.5) (with x replaced by z). The surface constant-head boundary condition, analogous to Equation (3.6) is

$$t \geq 0, \quad z = 0, \quad \psi = \psi_0 \tag{3.16a}$$

or

$$t \geq 0, \quad z = 0, \quad \theta = \theta_0 \tag{3.16b}$$

When the surface is ponded, we have in these equations $\psi_0 = 0$ or $\theta_o = \theta_s$. The surface flux boundary condition is written

$$t \geq 0, \quad z = 0, \quad v_0 = -K\left(\frac{d\psi}{dz} - 1 \right) \tag{3.17a}$$

or in terms of θ,

$$t \geq 0, \quad z = 0, \quad v_o = -D\frac{d\theta}{dz} + K \tag{3.17b}$$

The boundary condition (3.17) is termed here the flux boundary condition, and condition (3.16) may be called the constant head or constant potential boundary condition. Often, in terms as used for mathematics of diffusion, condition (3.16) is termed the "constant concentration" condition. It has also been called a "sudden ponding" condition.

For both Equations (3.2) and (3.15), the left side is often modified so that a single independent variable (ψ) is indicated, using the slope of the retention curve, and calling it the *specific moisture capacity* function, $d\theta/d\psi \equiv C_r(\psi)$:

$$C_r \frac{\partial \psi}{\partial t} = \frac{\partial}{\partial z}\left(K \frac{\partial \psi}{\partial z} - K \right) \tag{3.18}$$

Like the gravity-free case above, one may form a θ -based version of Equation (3.15) by substituting the soil diffusivity, and in this case one forms a convection-diffusion equation:

$$\frac{\partial \theta}{\partial t} = \frac{\partial}{\partial z}\left(D \frac{\partial \theta}{\partial z} - K(\theta) \right) \tag{3.19}$$

Philip's Series Solution

Philip (1957a) approached the solution of Equation (3.19) subject to conditions (3.5) and (3.16b) as a perturbation of the corresponding solution of the absorption equation (3.2). With successive transformations and approximations, he showed that the solution may be expressed as a power series in $t^{1/2}$:

$$z(\theta,t) = A_1 t^{1/2} + A_2 t + A_3 t^{3/2} + A_4 t^2 + ... \tag{3.20}$$

which series is carried to the accuracy desired. Coefficient A_1 is the Boltzmann variable φ, whose value is defined in Equation (3.9). The coefficients of the other terms are defined similarly but with increasing recursive complexity. For example, the value of A_2 is given by an implicit solution to the following [Philip, 1969]:

$$\int_{\theta_i}^{\theta} A_2 \, d\theta = \frac{D \dfrac{dA_2}{d\theta}}{\left(\dfrac{d A_1}{d\theta} \right)^2} + (K - K_i) \tag{3.21}$$

The method is described in more detail in [Philip,1969]; it is in general not practical for hydrologic application. However, a truncation of this series is commonly used, and appears in hydrologic research literature. The first two terms of the series are used as an infiltration equation. Instead of the definition of Equation (3.21), however, K_s - K_i is used for A_2, an approximation suggested by Philip [1957c] as being valid for both short and long times. Unfortunately, however, the approximation is quite biased at intermediate times, prompting Philip [1987] to

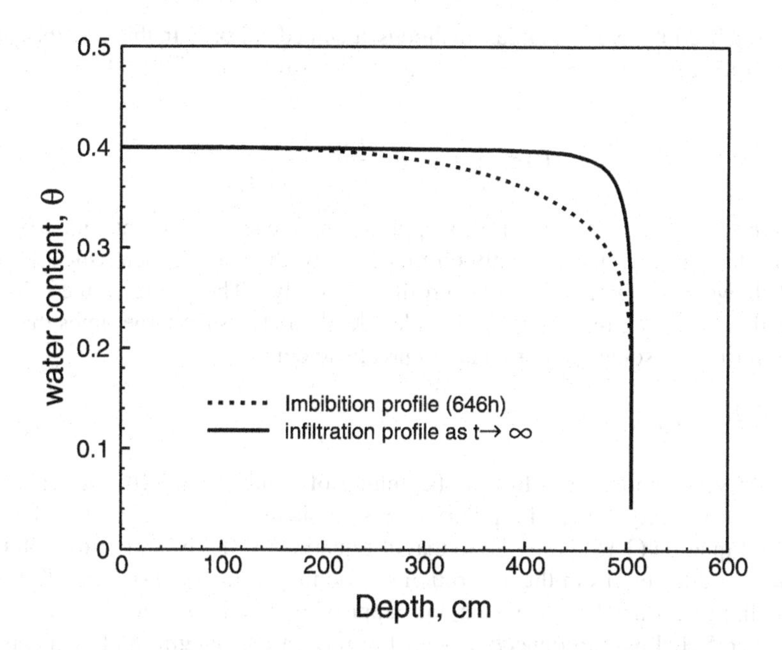

Figure 3.2. Comparison of the imbibition wetting profile with the infiltration profile at near infinite time (Equation 3.23) highlights the effect of gravity on the wetting profile.

suggest remedial corrective methods. This truncated series approximation is dealt with in more detail in further chapters.

Large-Time Solution

Equations (3.15) or (3.19), for either of the boundary conditions used above, describe a wetting "wave" moving away from the upper boundary down into the soil. For either condition, the wetting "wave" approaches a calculable shape at very long times, called by Philip the "profile at infinity". Philip [1957b] demonstrated mathematically that the long-time asymptotic value for the infiltration rate is $K_o = K(\theta_o)$, where θ_o or $\theta(\psi_o)$ is the imposed surface condition in the case of Eq. (3.16), or is K_s when $v_o \geq K_s$ in condition (3.17). If $v_o < K_s$, then the asymptotic flux value is $v_o = K_o - K_i$. At very large times, the moving wave approaches pure translation, and so at every point the gravitational and diffusive fluxes sum to the same constant velocity, *u*:

$$u = \frac{K_o - K_i}{\theta_o - \theta_i} \tag{3.22}$$

where subscript i refers to initial conditions, assumed uniform. In this case the equation (3.15) can be reduced to solve for a relative wave shape, $\zeta(\theta)$ [Philip, 1969]:

$$\zeta(\theta) = \int_{\theta}^{\theta_o - \varepsilon} \frac{D\, d\theta}{u(\theta - \theta_i) - (K - K_i)} \tag{3.23}$$

where ε is a small offset to prevent approaching the singular point, and $\zeta(\theta_0 - \varepsilon) \equiv 0$. This solution depends only on the shape of the soil characteristics. Figure 3.2 shows an example of such a "profile at infinity". The depths in the solution and this figure are relative to some index depth, and at such large times the profile must be at some great, probably unrealistic depth.

Short Time Solution

At very short times after the beginning of conditions (3.16), the capillary gradients are very high and capillary flux is dominant. Thus the short time behavior of Equation (3.15) is similar to that of Equation (3.2). The short time solution is in fact different from the absorption solution by a small constant at all times. The limiting time before which this approximation is useful is explored in Chapter 5, below, in connection with the infiltration integral and concepts of scaling. It depends, in short, on the hydraulic properties of the soil.

Analytical Solution

A particular sequence of transformations has been developed [Rogers *et al.*, 1983; Broadbridge and White, 1988; Sander *et al.*, 1988], subject to particular requirements on the form of the relations $D(\theta)$ and $K(\theta)$, that allows transformation of Equation (3.19) into the form of the Burger's Equation (see below), and thence solution by further transforms. In the following Chapter 4, the nature and potential of such analytic solutions will be explored. While the soil properties under the solution requirements can be made generally realistic, they cannot necessarily be fitted to an arbitrary set of measured soil hydraulic data. As shown below, the solution functions are complex and require numerical evaluation of transcendental functions. The solution is nevertheless quite useful at least as a tool by which to judge the quality of numerical methods.

MATHEMATICAL APPROXIMATIONS FOR SOIL WATER FLOW

Approximations to Eq. (3.15)

As indicated above, for short times after imposition of wetting at the soil surface, the gravitational term in Eq. (3.15) may be ignored, and a purely capillary diffusion solution used. The time period for which this approximation holds usefully is discussed below in the context of scaling. This approximation is rather underutilized in hydrology and in irrigation, for example, yet it allows significant simplification, relatively simple analytic equations, and use of physically-related soil parameters for calculations for irrigation intake, for example [Smith, 1999].

Burgers's Equation Approximation

This approximation to Richards' equation is of importance mostly because of it's analytic solvability. The θ-based form of Darcy's law [see Equation (3.17b)] is modified, retaining a special form of nonlinearity, such that flux is described by a constant D plus a second order dependence on θ:

$$v = -D_o \frac{\partial \theta}{\partial z} + A\theta^2 + B\theta + C \tag{3.24}$$

in which constants A, B, and C are arbitrary. We will employ this approximation as a transformation in the solution of the next chapter.

Kinematic Wave Approximation

As indicated above, for short times after the soil is wetted the gravitational terms may be ignored in comparison with flow due to the very large capillary potential gradients. For other very different cases, especially deep profiles with longer distances to a lower boundary, the capillary gradients have subsided, and purely gravitational flow may be assumed for certain flow problems and/or soils. This is often reasonable for lighter textured soils, because the major capillary potential gradients will resolve themselves in the upper part of the soil, and the slow downward seepage at greater depths will be characterized by gravitational fluxes. The equation that results is a kinematic wave equation. The fundamental assumption is that flux is a differentiable function of water content, θ. Continuity is again described by Equation (3.1), but now flux is described by the hydraulic characteristic relating hydraulic conductivity to water content. For the differentiable relation of v to θ we use a form of Equation (2.17):

$$v = K_s \Theta_e^{\varepsilon} \tag{3.25}$$

where ε is a soil parameter, and Θ_e is scaled water content defined in Equation (2.5). This is a reasonable description of observed behavior in many soils at longer time scales. These two relations are combined into the kinematic wave equation

$$\frac{d\theta}{dt}+\left[\varepsilon K_s \Theta_e^{\varepsilon-1}\frac{d\Theta_e}{d\theta}\right]\frac{d\theta}{dz}=j \tag{3.26}$$

The term in brackets is the characteristic wave velocity, $dv/d\theta = u_c$, the differential from Equation (3.25), and j is local gain/loss rate as in equation (3.1). This equation may be treated like the kinematic wave equations for surface water flow [Smith, 1983; Charbeneau, 1984], and has similar wave and shock properties, with $\varepsilon >> 1$. At the lower face of a soil wetting wave, with $d\theta/dz < 1$, the wave will steepen, since $du_c/d\theta > 0$, and in a short time will form a *shock*, or discontinuity in flux, when the wave front $d\theta/dz$ approaches $-\infty$. In this regard the shape of the long term profile, Figure 3.2, shows just such steepening. Using the flux relation of Equation (3.25), such kinematic shock front velocity u_s may be described as:

$$u_s(\theta_u,\theta_l)=\frac{K(\theta_u)-K(\theta_l)}{\theta_u-\theta_l}=K_s\frac{\Theta_{eu}^{\varepsilon}-\Theta_{el}^{\varepsilon}}{\theta_u-\theta_l};\quad \theta_u>\theta_l \tag{3.27}$$

Subscripts u and l refer to upper and lower, respectively. Note the similarity to the translation velocity of the wave at infinity, Equation (3.22). The difference in approach is that in the kinematic treatment, such advancing waves have step front shapes. The mechanics of routing and merging and decomposition of the waves formed from a variable input $v(t)$ at the soil surface was presented by Smith (1983). Because of the monotonic nature of the $v(\theta)$ relation, Equation (3.25), larger θ shocks will always tend to overtake and absorb smaller ones. Simple kinematic soil water waves also attenuate because characteristic velocities at the "back" (upward side) of the wave travel faster than the shock velocity. Figure 3.3 illustrates the relations between characteristic velocities and shock velocities. The chord slope of the $K(\theta)$ relation is the shock velocity, and the tangent $dK/d\theta$ at any θ is the characteristic velocity of that θ value. Thus the upper θ value of a shock will be overtaken by characteristics from above, and attenuation and elongation will result. Some of the relations involved in kinematic attenuation of a simple soil wetting wave are presented in the Appendix.

APPROXIMATIONS FOR SOIL HYDRAULIC CHARACTERISTICS

Delta-Function Diffusivity

Solutions for both horizontal and vertical soil water flow from a boundary are considerably simplified by assuming that water advances from the boundary as a simple square wave. This assumption relates more to the solution of the infiltration equation in subsequent chapters than the Richards' unsaturated flow equation, but the approximation is a basic one.

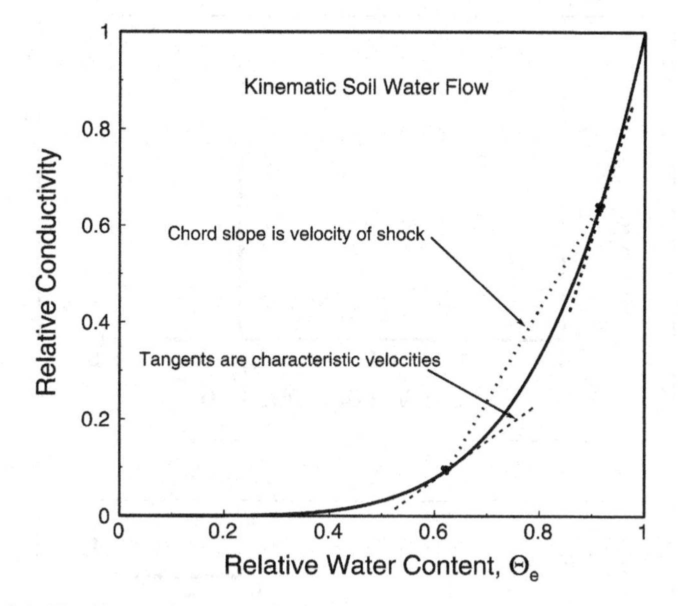

Figure 3.3. The kinematic wave soil water flow approximation is related to the relation $K(\theta)$, and there is a graphical interpretation, shown here, for the characteristic θ velocities and the $\Delta\theta$ shock velocities.

If the solution $\varphi(\theta)$ for constant $\theta = \theta_o$ boundary conditions shown in Figure 3.1 is treated as a square wave, it implies a single valued $\varphi = \varphi_D$. Further, from the definition in Equation (3.12), sorptivity must be a constant defined as

$$S_D = \varphi_D\,(\theta_o - \theta_i). \tag{3.28}$$

Since for this assumption all the change in θ occurs in the infinitesimal region near θ_o, it is evident from inspection that the absorption solution given in Equation (3.10) can be reduced, to define a diffusivity that is concentrated in the region near θ_o, and can be written (as Philip [1973] showed)

$$D_* = \frac{S_D^{\;2}}{2(\theta_o - \theta_i)}\,\delta(\theta_o - \theta) \tag{3.29}$$

in which $\delta(-)$ is the Dirac delta function, indicating here that D has θ-integral values only at $\theta = \theta_o$. Figures 3.4a and b illustrate the implications of this assumption in terms of the soil hydraulic characteristics $D(\theta)$ and $K(\psi)$. By the definition of D, the conductivity K must be equal to K_s as ψ is reduced from 0 to the value of ψ_D, where $\psi_D = -D_*/K_s$.

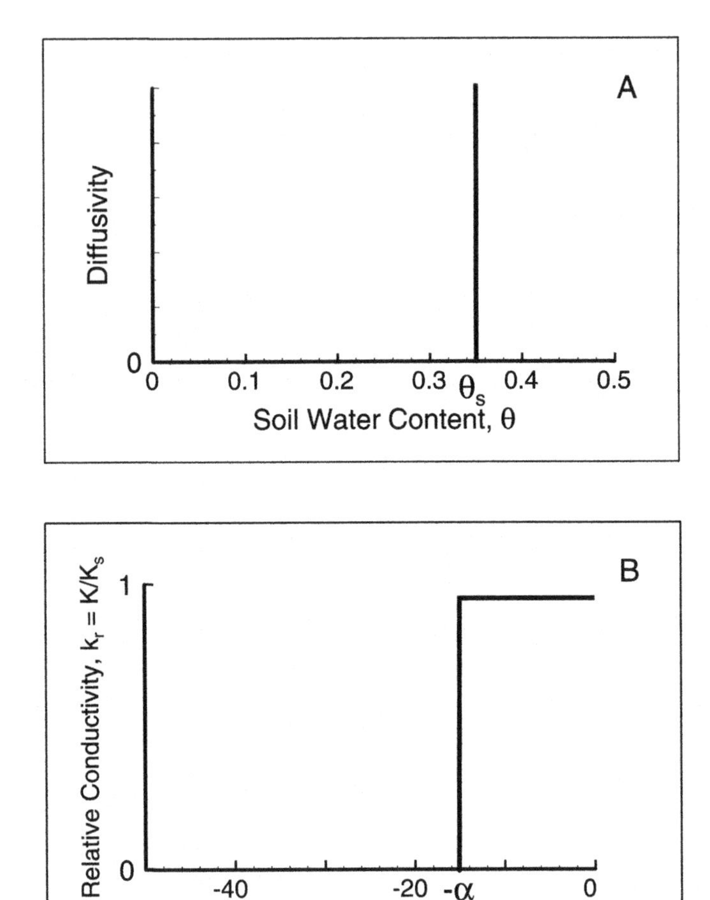

Figure 3.4. The assumption of a delta-function diffusivity concentrates all diffusivity at one water content as shown in A, and assumes that the soil water conductivity is a step function at a soil water potential of $-\alpha$ as shown in B.

It follows from Equations (3.14) and (3.28) that in a delta diffusivity soil the advancing profile is a step wave moving as $t^{1/2}$. This solution, with the addition of the gravity flow term, is consistent with the Green and Ampt [1911] solution for infiltration, and we revisit this assumption in Chapter 5. It is important because it represents one extreme but not unrealistic assumption that forms a bound for the behavior of real soils.

Exponential $K(\psi)$

Another assumption on the form of the soil characteristics that has been mathematically useful was presented by Philip[1969] as a "quasilinear" solution. Assuming that

$$K(\psi) = K_s e^{\alpha\psi} \tag{3.30}$$

and also employing the transformation for flux potential, Equation (2.8),

$$\phi(\psi) \equiv \int_{\psi_i}^{\psi} K d\psi \tag{3.31}$$

allows linearization of some two- and three-dimensional forms of the unsaturated flow equation, at the same time preserving at least an approximation to the important nonlinear nature of the soil characteristics. The key property of Equation (3.30) is that $dK/d\psi$ is a constant, and Equation (3.31) is important because $d\phi/d\psi$ is K. The approximation (3.30) is useful also in integrating the infiltration equation, demonstrated in Chapter 5. As a description of real soils, however, in general [or (3.25)], Equation (3.30) describes a K which decreases far too rapidly at larger absolute values of ψ.

The Analytic Solution Soils

The analytic solution indicated above, using a set of particular transformations, depends on a particular but not unrealistic form for the description of the soil characteristics to enable the transformations. The next chapter is devoted to this solution and its assumptions and will not be outlined here. The value of such a solution is not primarily for application to practical soil water flow problems. An analytic solution is an important means to test the accuracy of numerical solvers for Equations (3.15) or (3.19) that can then use more realistic boundary and soil properties. This approach also offers the means to look in detail at the relation of the solution properties to the soil water characteristic functions [White and Broadbridge, 1988]. Whatever the limitations of these functional forms for the soil characteristics, they better represent the nonlinearities of real soils than any others known to date which allow analytic solution.

SUMMARY

Both Darcy's law and the basic one-dimensional transient flow equations for unsaturated soil water flow can be written with either water pressure head or water content as the dependent variable. Flow from a boundary in response to a pressure gradient without the influence of gravity is called absorption or imbibition. Infiltration is absorption plus gravitational flow (downward). The infiltration equation is of the convection - diffusion type, meaning that it describes the movement of water due to diffusive as well as convective (gravitational) gradients. Diffusive flow of water in unsaturated soil is complicated by the highly nonlinear nature of the equivalent "diffusivity", and convective flow of soil water is also highly nonlinear through the relation of K to θ or ψ. Several approximations have been made to facilitate mathematical treatment of the flow equations, and some are introduced here. Because of these nonlinearities, numerical methods are necessarily employed to solve Richards' equation. Only one reasonable nonlinear description of soil hydraulic properties has been discovered which allows for a solution of the infiltration equation, subject to steady rainfall conditions. In the next chapter we turn our attention to this solution of the infiltration equation subject to the associated description of soil hydraulic properties.

4

A Realistic Integrable Nonlinear Model for Transient Unsaturated Flow

Philip Broadbridge

Chair, Dept. of Mathematical Sciences, University of Delaware, Newark, Delaware

INTRODUCTION

As emphasised many times throughout this book, a central concern of infiltration theory is to solve boundary value problems involving the nonlinear time-dependent Richards unsaturated flow equation [Equation 3.19], which can be written

$$\frac{\partial \theta}{\partial t} = \nabla \cdot \left(D(\theta)\nabla\theta\right) - K'\frac{\partial \theta}{\partial z} \tag{4.1}$$

In these boundary value problems, the solutions are required to satisfy ideal but meaningful boundary conditions. For many such problems, we must resort to approximate methods of numerical computer simulation. In these times, unlike in the days when the continuum models of unsaturated flow were first formulated, we have access to computing hardware and software that is usually capable of performing such tasks to a satisfactory level of accuracy. However, this does not mean that analytical solution methods are no longer important. Analytical solutions sometimes lead to simple functional relationships among measurable quantities that would be difficult to discern from numerical outputs alone. An example of this is Philip's infiltration series discussed previously in Chapter 3. Analytical solutions tend to be more efficient. They already express information on the dependence of the solution on system parameters and boundary data. For example, the dependence of ponding time on irrigation rate is expressed most succinctly from approximate and exact analytic solutions [Parlange and Smith, 1976; Clothier *et al.*, 1981; Broadbridge and White, 1987]. Similar information could be gleaned from numerical methods only by rerunning numerical simulation programs many times to cover a range of irrigation rates specified within the

Infiltration Theory for Hydrologic Applications
Water Resources Monograph 15

boundary conditions. Finally, numerical simulation programs that are sufficiently adaptable to solve a wide range of boundary value problems usually require a degree of sophistication that precludes *a-priori* error estimates. Ultimately, they need to be bench-tested against exact solutions of model problems.

In truth, there are very few exact time-dependent solutions to fundamental boundary value problems involving the nonlinear Richards equation. In two and three spatial dimensions, some solutions emerge from a systematic Lie symmetry classification [Edwards and Broadbridge, 1994]. However, while some of these satisfy plausible boundary conditions, we do not regard them as fundamental. For this reason, we are forced to focus attention again on unidirectional vertical flows. Henceforth, the gradient vector operator in Richards' equation will be replaced by the scalar operator $\alpha/\alpha z$. Most insight on unsaturated flow is expected to be gained from solutions that satisfy basic uniform initial conditions $\theta(\zeta,0) = \theta_i$ and boundary conditions of prescribed constant concentration, given in Equation (3.16) as

$$\theta\ (0,\tau)= \theta_0$$

or prescribed constant flux as in Equation (3.17):

$$v(0,t) = r = K\ (\theta_0).$$

In the non-ponding case $r = K(\theta_0) < K_s$, θ_0 is the asymptotic value of surface water concentration at large t. For either of these boundary conditions, the solution is known to agree asymptotically at large-t with a travelling wave $\theta = f\ (\phi)$, where ϕ is the d'Alembert variable z-Ut, with velocity $U= (K_0 - K_i)/(\theta_0 - \theta_i)$. The shape of the wave, which Philip [1957] called "the profile at infinity", is the solution of Equation (3.23), and is also the travelling wave solution of the ordinary differential equation

$$D(f)f\,''(\phi)+D'(f)(f\,')2 +[U-K'(f)]f\,' \ = \ 0$$

with $f(\phi) \rightarrow \theta_i\ ,\theta_0$ as $\phi \rightarrow \infty\ , -\infty$.

However at early times the travelling wave solution does not satisfy relevant boundary conditions at z =0.

The constant flux boundary condition (3.17a) in ψ may also be written in θ (3.17b) as

$$v(z,t) = K(\theta) - D(\theta)\frac{\partial\theta}{\partial z} = r \qquad \text{at z=0.} \tag{4.2}$$

This is a complicated nonlinear constraint. For a long time, the problem of constructing an exact solution with constant flux boundary condition remained unsolved. For this purpose, we now have available a variety of integrable forms of the Richards equation.

INTEGRABLE FORMS OF RICHARDS' EQUATION

The most common method for obtaining exact solutions of nonlinear partial differential equations is the classical method of Lie symmetry reductions [e.g. Bluman and Kumei, 1985]. However, the prescribed flux boundary condition is unlikely to be invariant under any foreseeable symmetry transformation of any nonlinear version of Richards' equation. There is more hope of solving this problem using an integrable nonlinear version of Richards' equation that can be transformed to a linear equation by a change of variable. After transforming to a linear equation, we would have linear transform methods at our disposal. Fortunately, the class of quasilinear parabolic partial differential equations contains a rich variety of these linearisable models. Up to a contact transformation (a self-consistent change of variable that allows a new dependent variable to be defined in terms of old variables and their first derivatives), the integrable second order evolution equations belong to one of four inequivalent classes [Svinolupov, 1985]. One set of forms for these classes is given below, for arbitrary variable u:

(i) the linear class equivalent to $\dfrac{\partial u}{\partial t} = \dfrac{\partial^2 u}{\partial x^2} + g(x)u$ (g arbitrary)

and including the linear model for (4.1) [e.g. Philip, 1969; Braester, 1973]

$$\frac{\partial \theta}{\partial t} = D\frac{\partial^2 \theta}{\partial z^2} - U\frac{\partial u}{\partial z} \tag{4.3}$$

with D and U constant,

(ii) the Burgers class equivalent to $\dfrac{\partial u}{\partial t} = \dfrac{\partial^2 u}{\partial x^2} + 2u\dfrac{\partial u}{\partial x} + g(x)$ (g arbitrary)

and including the weakly nonlinear model for (4.1) [Clothier *et al.*, 1981; Broadbridge, 1999]

$$\frac{\partial \theta}{\partial t} = D\frac{\partial^2 \theta}{\partial z^2} - 2\sigma(\theta - \theta_n)\frac{\partial \theta}{\partial z}, \tag{4.4}$$

with D, σ and θ_n constant,

(iii) the Fujita class, equivalent to $\frac{\partial u}{\partial t} = \frac{\partial}{\partial x}\left[u^{-2}\frac{\partial u}{\partial x}\right]$

and including the realistic model Richards equation [Broadbridge and White, 1987, 1988; Sander *et al.*, 1988]

$$\frac{\partial \theta}{\partial t} = \frac{\partial}{\partial z}\left[\frac{a}{(b-\theta)^2}\frac{\partial \theta}{\partial z}\right] + \left[\gamma - \frac{\lambda}{2(b-\theta)^2}\right]\frac{\partial \theta}{\partial z}, \tag{4.5}$$

with a,b,γ and λ constant, and

(iv) the Freeman- Satsuma [1989] class, equivalent to

$$\frac{\partial u}{\partial t} = \frac{\partial}{\partial x}\left[u^{-2}\frac{\partial u}{\partial x}\right] - 2$$

and including the Richards equation with distributed plant root absorption term [Broadbridge and Rogers, 1993]

$$\frac{\partial \theta}{\partial t} = \frac{\partial}{\partial z}\left[\frac{a}{(b-\theta)^2}\frac{\partial \theta}{\partial z}\right] - \upsilon\frac{a}{(b-\theta)^2}\frac{\partial \theta}{\partial z} - Qe^{-\upsilon z} \tag{4.6}$$

with a,b, υ and Q constant.

The linear equation (4.3) assumes a linear function for conductivity $K(\theta)$. However, measured conductivity functions $K(\theta)$ are concave upwards, a property that ensures that the large-t travelling wave solution is stable. In practice, a linear interpolating function $K(\theta)$ between two widely separated measured pairs (θ_n, K_n) and (θ_0, K_0) grossly overestimates the conductivity at intermediate values of θ. This leads to an exaggeration of the effect of gravity and a severe over-estimate of the time to ponding during steady irrigation [Broadbridge and White, 1987].

Burgers Equation

Since the linear diffusion-convection equation is inadequate for modeling unsaturated flow, the discovery in the 1950's of integrable nonlinear parabolic equations proved to be a pivotal development. Motivated by the Burgers [1948] study of a simple prototype for nonlinear fluid dynamics, Hopf [1950] and Cole [1951] transformed Burgers' equation, the source-free canonical form of class (ii), to the linear heat equation (with some little-known precedent in a student exercise set by Forsyth [1906]). In an unrelated study of heat conduction in metals, Storm [1951] transformed the class (iii) nonlinear diffusion equation with inverse square nonlinear diffusivity to the linear heat equation. Using more complicated transformations, Fujita [1951, 1952, 1953] constructed exact solutions to nonlinear diffusion equations with reciprocal quadratic diffusivity, satisfying Dirichlet constant-concentration boundary and initial conditions. Knight and Philip[1974] improved Storm's transformation method to obtain solutions to the Cauchy redistribution problem.

The Burgers equation has a nonlinear convective term that stabilizes the travelling wave solution, which in the context of unsaturated flow, exhibits a wetting front that smoothly switches the system from its dry initial conditions to its wet final conditions. This behavior is observed in the solutions with constant-flux boundary conditions and constant-concentration boundary conditions, as they agree with the travelling wave solution asymptotically at large-t. This equation has been solved exactly for a variety of boundary and initial conditions, including prescribed constant flux conditions and prescribed concentration conditions [Benton and Platzman, 1972]. It has been used as a model for field soils, within which macropores may hold water at near-zero potential energy, resulting in a smoother water retention-potential curve [Clothier et al., 1981]. It is useful for predicting the time to incipient ponding, under conditions of constant or variable irrigation rate [Clothier *et al.*, 1981; Broadbridge and White, 1987; Broadbridge and Stewart, 1996]. However, in the modeling of highly nonlinear soils, including repacked laboratory soils, its linear diffusion term is a major deficiency. This cannot capture the inflection point in the wetting front that is observed in water content profiles of nonlinear soils even at early infiltration times. This nonlinearity is effectively restored in the class (iii) diffusion equation studied by Storm [1951], Fujita [1952], and Knight and Philip [1974]:

$$\frac{\partial \theta}{\partial t} = \frac{\partial}{\partial x}\left[\frac{a}{(b-\theta)^2}\frac{\partial \theta}{\partial x}\right] = \frac{a}{(b-\theta)^2}\frac{\partial^2 \theta}{\partial x^2} + \frac{2a}{(b-\theta)^3}\left(\frac{\partial \theta}{\partial x}\right)^2 \qquad (4.7)$$

Note that this equation has no convective term at all. For the purposes of soil hydrology, this can be used only when gravity can be neglected; in space away from planets, in controlled horizontally constrained flow or for vertical flow at early times. Later we will recount some of the lines of thought that allowed useful convective terms to be added to Equation (4.7). To do this, we first return to the Storm transformation, which was the first linearising transformation to be applied to Equation (4.7). However, we keep in mind that any linearisable equation has many alternative linearising transformations, and these might be useful for obtaining different classes of solutions.

SOLUTION OF NONLINEAR DIFFUSION EQUATION

First, we reduce the number of terms in Equation (4.7) by applying the Kirchhoff [1894] transformation

$$\mu = \int D(\theta)d\theta = \frac{a}{(b-\theta)} \tag{4.8}$$

to obtain from Equation (4.7):

$$\frac{\partial \mu}{\partial t} = a^{-1}\mu^2 \frac{\partial^2 \mu}{\partial x^2} \tag{4.9}$$

Now we apply the Storm transformation, in the notation of Knight and Philip [1974];

$$\chi = \int_0^x D^{-1/2}[\theta(y,t)]dy$$

$$= a^{1/2}\int_0^x \mu(y,t)dy$$

$$= a^{-1/2}\int (b-\theta(y,t))dy \tag{4.10}$$

and $\tau = t$.

The coordinate transformation from (x,t) to (χ, τ) must be applied to Equation (4.9) with care. Although $\tau = t$, the operators $\partial/\partial\tau$ and $\partial/\partial t$ are not equal:

$$\frac{\partial \mu}{\partial t} = \frac{\partial \mu}{\partial \tau} + a^{-1/2}\frac{\partial \mu}{\partial \chi}\left[-a^{1/2}\mu^{-1}\frac{\partial \mu}{\partial \chi} - v(0,\tau)\right],$$

where v(x,t) is the Darcian volumetric water flux,

$$v = -D(\theta)\frac{\partial \theta}{\partial x} = a^{1/2}\mu^{-1}\frac{\partial \mu}{\partial \chi}$$

In terms of the new variables, Equation (4.9) is

$$\frac{\partial \mu}{\partial \tau} = \frac{\partial^2 \mu}{\partial \chi^2} + a^{-1/2} v(0,\tau)\frac{\partial \mu}{\partial \chi} \tag{4.11}$$

This is now a linear equation. Although the final term takes the appearance of a convection term, this is an artifact of the transformations, as the equivalent equation (4.7) contains no convection term. For constant prescribed-flux boundary conditions on (4.7),

$$v(0,\tau) = r \text{ (constant)},$$

resulting in constant coefficients within Equation (4.11), making it amenable to standard linear transform methods [e.g Carslaw and Jaeger, 1959]. For uniform-concentration initial conditions, and constant-concentration boundary conditions,

$$\mu(\chi,0) = \mu_i = \frac{a}{(b-\theta_i)} \tag{4.12a}$$

$$\mu(0,\tau) = \mu_0 = \frac{a}{(b-\theta_0)} \tag{4.12b}$$

and the surface flux is known to be

$$v(0,\tau) = S\,\tau^{-1/2}, \tag{4.13}$$

where S is the sorptivity (Equation 3.12). This enables us to solve the boundary value problem exactly by similarity reduction. Importantly, this yields an explicit relationship between sorptivity, initial water content, boundary water content and the model diffusivity parameters a and b.

Assuming conditions (4.12), Equation (4.11) is invariant under the rescaling

$$\bar{\chi} = e^{\varepsilon}\chi \; ; \; \bar{\tau} = e^{2\varepsilon}\tau \quad (\varepsilon \text{ arbitrary}),$$

This transformation tacitly assumes $\bar{\mu} = \mu$. That is, μ does not change, or μ is an invariant. Another invariant is $Y = \chi/\sqrt{\tau}$ since

$$\frac{\bar{\chi}}{\sqrt{\bar{\tau}}} = \frac{\chi}{\sqrt{\tau}} \ .$$

Assuming an invariant solution $\mu = f(Y)$, Equation (4.11) reduces to an ordinary differential equation

$$\frac{d\rho}{dY} = -\frac{1}{2}\left(a^{-1/2}S + Y\right)\rho \qquad where\ \rho = f'(Y)$$

The solution satisfying (4.12) is

$$\mu = \mu_i + \frac{\mu_0 - \mu_i}{erfc\left(\frac{a^{-1/2}S}{2}\right)} erfc\left(\frac{a^{-1/2}S + Y}{2}\right) \tag{4.14}$$

So far, we have not fully specified the sorptivity, S, even though it appears in the solution. For consistency, the solution (4.14) must imply a flux for Equation(4.7) that agrees with Equation (4.13). From Equation (4.13), we deduce a transcendental equation for S, $S = (a/h(C))^{1/2}$, where h(C) is defined by

$$\sqrt{\pi}\left(\frac{h^{-1/2}}{2}\right) erc\left(\frac{h^{-1/2}}{2}\right) = \frac{1}{C} \tag{4.15}$$

with the *erc* function defined by

$$erc(x) = exp\,(x^2)\ erfc\,(x)\ , \tag{4.16}$$

and C is the nonlinearity parameter

$$C = \frac{b - \theta_i}{\theta_0 - \theta_i} \tag{4.17}$$

which is just above 1 for highly nonlinear soils, and far above 1 for almost-linear soils.

The nonlinearity parameter determines the relative change in diffusivity over the range of water content θ_i to θ_0, as

$$\frac{D(\theta_0)-D(\theta_i)}{D(\theta_i)}=\frac{C^2}{(C-1)^2}-1$$

In practice, we have found it convenient to approximately invert Equation (4.14) by taking the explicit form [White and Broadbridge, 1988]

$$h(C) = C(C-1)[\pi (C-1) + B]/[4(C - 1) + 2B] \pm 1\% \qquad (4.18)$$

with B=1.46147 .

As the nonlinearity parameter C varies over the entire range from 1 (extreme nonlinearity) to ∞ (linear diffusion), the parameter h(C)/ [C(C-1)] varies within the narrow range from ½ to $\pi/4$.

Note that this parameter divided by squared water content range is the ratio of integral diffusivity to squared sorptivity,

$$\frac{hC}{C(C-1)}=\frac{(\theta_0-\theta_i)^2}{S^2}\overline{D}=\frac{1}{S^2}(\theta_0-\theta_i)\int_{\theta_i}^{\theta_0}D(\theta)d\theta \ . \qquad (4.19)$$

Its range of values for other nonlinear diffusion models, and its relationship to other soil parameters, have been discussed by Warrick and Broadbridge [1991].

REQUIRED FORMS FOR SOIL CHARACTERISTICS

Given that Equation (4.7) may be integrated, then it is immediately obvious that we can incorporate a linear convection term, since the equation

$$\frac{\partial\theta}{\partial t}=\frac{\partial}{\partial z}\left[\frac{a}{(b-\theta)^2}\frac{\partial\theta}{\partial z}\right]-U\frac{\partial\theta}{\partial z} \qquad (4.20)$$

is equivalent to (4.7) by use of a moving coordinate z = x+Ut.

In the context of unsaturated flow modeling, a linear convection term in (4.20), as applied by O'Kane *et al.* [1981] is an improvement on no convection term. However, Equation (4.5), which has an additional nonlinear convection term, has proven to be more versatile for various fields of application.

Let us consider a general nonlinear convection term appended to Equation (4.7),

$$\frac{\partial\theta}{\partial t}=\frac{\partial}{\partial z}\left[\frac{a}{(b-\theta)^2}\frac{\partial\theta}{\partial z}\right]-K'(\theta)\frac{\partial\theta}{\partial z} \qquad (4.21)$$

If, as before, we apply the Kirchhoff transformation (4.8) followed by the Storm transformation (4.10), we find that we achieve a linear diffusion-convection equation, similar to Equation (4.11) only if $K(\theta)$ is a linear function. Next, we ask when these transformations can result in the linearisable Burgers equation,

$$\frac{\partial\mu}{\partial t}=\frac{\partial^2\mu}{\partial\chi^2}+(m_1\mu+m_2)\frac{\partial\mu}{\partial\chi}\ , \qquad (4.22)$$

with m_1 and m_2 constant. The answer is that $K(\theta)$ may take the general form

$$K(\theta) = \beta + \gamma(b-\theta) + \frac{\lambda}{2(b-\theta)}, \qquad (4.23)$$

with b, β, γ and λ constant.

The integrability of Equation (4.5) is apparent from the independent works of Fokas and Yortsos [1982], who applied it to water-oil displacement, and of Rosen [1982] who applied it to transport of solute with adsorption-solution equilibrium. Each of these works made use of Equation (4.5) with $\gamma = 0$. Rogers et al. [1983] extended the model of Fokas and Yortsos [1982] to incorporate gravity by taking γ non-zero.

For the purposes of modeling unsaturated flow, the singularity in $K(\theta)$ at $\theta = b$ will not cause any problem, provided b is chosen to be larger than the maximum water content θ_0. Usually, b is taken to be larger than θ_S . We must choose $\gamma > 0$, so that the function $K(\theta)$ has a positive second derivative, a significant improvement on the linear model. If $\gamma > 0$, the function $K(\theta)$ has a single local and global minimum at

$$\theta = \theta_n = b - \sqrt{\frac{\lambda}{2\gamma}} \qquad (4.24)$$

Therefore, if we wish $K(\theta)$ to be an increasing function in the domain of interest, we must take $\theta_n \leq \theta_i$. For comparison with the normalization of Equation (2.13), n is comparable to θ_r. In the formulation of Broadbridge and White [1988] it was assumed that the initial water content was sufficiently low that $K'(\theta_i)$ could safely be assumed to be zero. This is equivalent to assuming $\theta_n = \theta_i$. This leads to simplifications in the mathematical solutions, the convenience of which usually more than compensates for minor errors in the model hydraulic functions. However, we envisage some situations in which the case $\theta_i > \theta_n$ might need to be considered. For example, we might wish to predict the time to incipient ponding after a soil has previously been partially wetted to a uniform initial water content in the region of which $K(\theta)$ is clearly increasing. As explained by White and Broadbridge [1988], when comparing outcomes from different initial water contents, according to Equation (4.17), different values must be used for the nonlinearity parameter C. When such a distinction is necessary, we shall denote

$$C_i = \frac{b - \theta_i}{\theta_0 - \theta_i} ,$$

and

$$C_n = \frac{b - \theta_n}{\theta_0 - \theta_n} .$$

Note that for any of these definitions for C, in the limit as $C \to \infty$ the analytically solvable form of Richards' equation reduces to the weakly nonlinear Burgers' equation, whereas in the limit as $C \to 1$ it reduces to a highly nonlinear model with delta function diffusivity. This nonlinear extreme is not identical with the Green-Ampt model. For example, it leads to a different and more accurate prediction of the time to ponding during irrigation [Broadbridge and White, 1987]. The notion of this limiting approach has caused some confusion. Equation (4.5) has four free parameters. There are many curves in the four dimensional parameter space that have b approaching infinity. We choose a curve along which directly measurable properties such as sorptivity S are held constant [Broadbridge and White, 1988; White and Broadbridge, 1988]. Along such a curve, the parameter $a = h(C)\, S^2$ cannot be constant as C varies. The curve with C increasing and constant a is of no physical relevance, as it passes through regions where S takes ridiculous values.

Here, we choose to express hydraulic functions in terms of C_n and we define

$$\Theta = \frac{\theta - \theta_n}{\theta_0 - \theta_n} ,$$

similar to Equation (2.13). This convention allows us the most direct comparison with the solution of Broadbridge and White [1988]. However we allow more general initial conditions

$$\theta_i \geq \theta_n ,$$

so that the initial value of Θ may be greater than zero. Since the initial condition is not necessarily at the minimum conductivity, the mathematical solution is a little more complicated.

In keeping with the principles of similitude [e.g. Birkhoff, 1930; Fulford and Broadbridge, 2001], the shape of the water content profile must be expressible in terms of dimensionless coordinates, and it must be determined from dimensionless parameters of the boundary value problem. For our length scale, we choose the capillary length λ_s over the range of water content from θ_n to θ_s :

$$\lambda_s = \frac{1}{K_s - K_n} \int_{\psi_n}^{0} K(\psi) d\psi$$

$$= \frac{1}{K_s - K_n} \int_{\theta_n}^{\theta_s} D(\theta) d\theta$$

$$= \frac{h(C_n) S^2}{C_n (C_n - 1)(\theta_s - \theta_n)(K_s - K_n)} \tag{4.25}$$

This conforms to Philip's [1985] generalization of the Gardner [1958] parameter $1/\alpha$ appearing in the often-used exponential model $K = K_s \exp(\alpha\Psi)$ (Equation 3.30). Denoting by U_n the maximum travelling wave speed $(K_s - K_n)/(\theta_s - \theta_n)$, a natural time scale is

$$t_s = \lambda_s / U_n = \frac{h(C_n) S^2}{C_n (C_n - 1)(K_s - K_n)^2} \tag{4.26}$$

Except for the factor $h(C)/[C(C-1)]$ which is typically 0.55 for strongly nonlinear soils [Warrick and Broadbridge, 1991], t_s is the gravity time scale [Philip, 1969].

Now we define dimensionless variables $z* = z/\lambda_s$, and $t_{s*} = t/t_s$, in terms of which the Richards equation is

$$\frac{\partial \Theta}{\partial t_{s*}} = \frac{\partial}{\partial z_*}\left(D_* \frac{\partial \theta}{\partial z_*} \right) - \frac{\partial K_*}{\partial z_*} \tag{4.27}$$

where $D_* = D/\bar{D} = C_n(C_n - 1)/(Cn - \Theta)^2$. $\bar{D}$ being the mean diffusivity

$$\overline{D} = \lambda_s^2 t_s^{-1} = \frac{h S^2}{(\theta_S - \theta_n)^2 C_n (C_{n-1})},$$

and

$$K_* = \frac{K - K_n}{K_S - K_n} = \frac{(C_n - 1)\Theta^2}{(C_n - \Theta)} \tag{4.28}$$

From Equation (4.28) and the definition of D (Equation 3.3), one can obtain the characteristic retention relation $\Theta(\psi)$ for the soil defined in this manner by parameter C_n [Broadbridge and White, 1988]. For negligible values of K_n, this becomes

$$\psi(\Theta) = -\lambda_s \frac{1-\Theta}{\Theta} - \frac{1}{C_n} \ln\left[\frac{C_n - \Theta}{\Theta(C_n - 1)}\right] \tag{4.29}$$

Figure 4.1 illustrates this relation for a range of C_n values, and the $K_*(\Theta)$ relation is illustrated in Figure 4.2. Combining these two, the relation of K_* to scaled ψ for a range of C_n is shown in Figure 4.3. These curves have equal integrals by virtue of being scaled on λ_s.

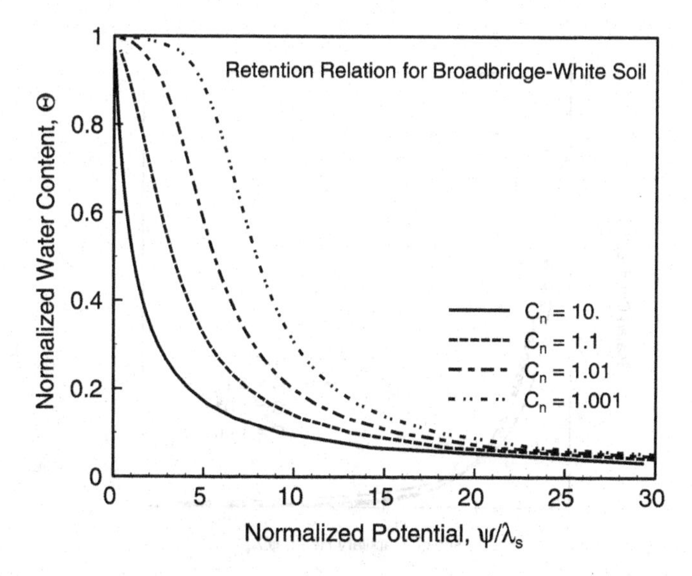

Figure 4.1. The retention relation, Equation (4.29), of the integrable soil characteristics varies markedly with the parameter C, but a quite realistic retention function is described.

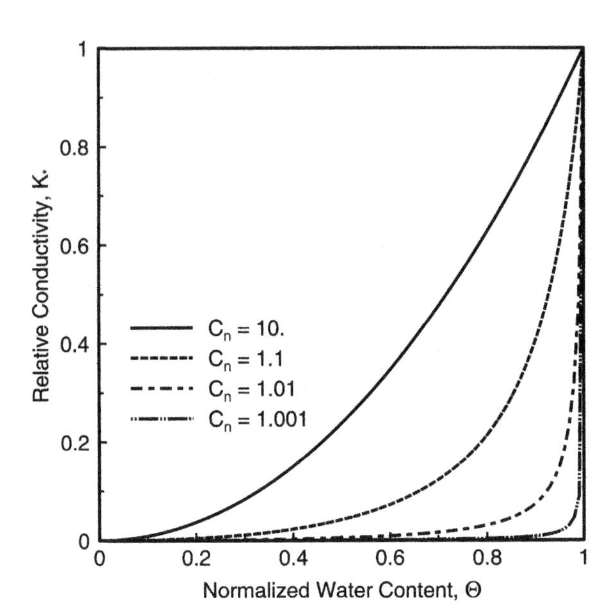

Figure 4.2. The conductivity expressed as a function of normalized Θ is shown for the same parameter C values as for Figure 4.1. The dramatic change in characteristic nonlinearity is easily seen here.

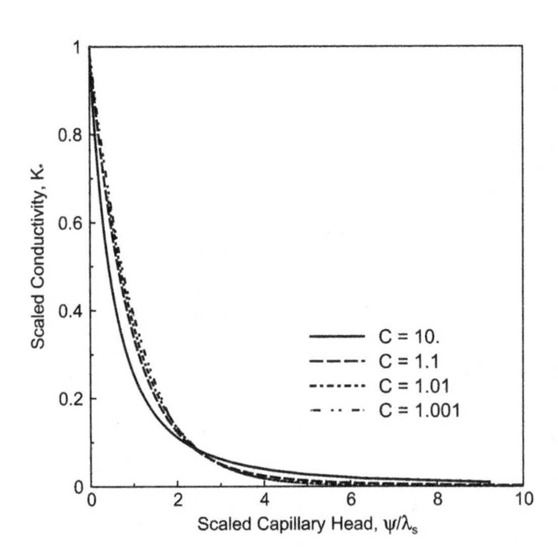

Figure 4.3. The relation of scaled K_* to capillary potential for the Broadbridge-White soil for several values of the nonlinearity parameter C.

SOLUTION OF THE INTEGRABLE FORM

The initial condition is

$$\Theta = \Theta_i \geq 0 \text{ at } t_{s*} = 0. \tag{4.30}$$

The boundary conditions are

$$v_*(z_*) = K_* - D_* \, \partial\Theta/\partial z_* = r_* \text{ at } z_* = 0 \tag{4.31}$$

and

$$\Theta \rightarrow \Theta_i \text{ as } z_* \rightarrow \infty.$$

Here, $r_* = (r - K_n)/(K_s - K_n)$. From here on, the relevant Kirchhoff variable will be taken to be $M = C_n(C_n - 1)/(C_n - \Theta)$, in terms of which the boundary value problem is

$$\frac{\partial M}{\partial t_{s*}} = \frac{1}{C_n(C_n - 1)} M^2 \frac{\partial^2 M}{\partial z_*^2} + \left[C_n - 1 - \frac{M^2}{C_n - 1}\right] \frac{\partial M}{\partial z_*}, \tag{4.32}$$

$$M = M_i = C_n(C_n - 1)/(C_n - \Theta_i) \quad \text{at } t_{s*} = 0,$$

$$\frac{C_n}{M}\left[(C_n - 1)^2 - 2(C_n - 1)M + M^2\right] \frac{\partial M}{\partial z_*} \quad \text{at } z_* = 0, \tag{4.33}$$

$$M \rightarrow M_i \text{ as } z_* \rightarrow \infty.$$

Now we apply the rescaled version of the Storm transformation,

$$Z = \int_0^{Z_*} D_*^{-1/2}\left[M(y_*, t_{s*})\right] dy_*$$

$$= \left[C_n(C_n - 1)\right]^{-1/2} \int_0^{Z_*} \left[C_n - \Theta(y_*, t_{s*})\right] dy_*$$

$$\left[C_n(C_n - 1)\right]^{1/2} \int_0^{Z_*} M^{-1} dy_* \tag{4.34}$$

$T = t_{s*}$.

This results in a form of Burgers' equation as the governing equation,

$$\frac{\partial M}{\partial T} = \frac{\partial^2 M}{\partial Z^2} - 2[C_n(C_n - 1)]^{1/2}\left[-1 - 2\rho + \frac{M}{C_n - 1}\right]\frac{\partial M}{\partial Z}, \qquad (4.35)$$

where $\rho = r_*/ 4C_n(C_n - 1)$.

The initial and boundary conditions are

$$M = M_i = C_n(C_n - 1)/(C_n - \Theta_i) \qquad T = 0,\ Z \geq 0$$

$$C_n M^2 - 2C_n(C_n - 1)[1+2\rho]M + C_n(C_n - 1)^2 - [C_n(C_n - 1)]^{1/2}\,\partial M/\partial Z = 0$$
at $z_* = 0$, (4.36)

$M \rightarrow M_i$ as $z_* \rightarrow \infty$.

After applying the Hopf-Cole transformation

$$-1-2\rho + M/(C_n - 1) = -u^{-1}u/\zeta \qquad (4.37)$$

to Equation (4.36), with $\zeta = [C_n(C_n - 1)]^{1/2} Z$, it is sufficient that u satisfies the linear diffusion equation

$$\frac{\partial u}{\partial \tau} = \frac{1}{4}\frac{\partial^2 u}{\partial \zeta^2} \qquad (4.38)$$

with $\tau = 4C_n(C_n - 1)\,T$.

Fortunately, the flux boundary condition (4.36) also simplifies to

$$\frac{\partial^2 u}{\partial \zeta^2} = 4\frac{\partial u}{\partial \tau} = 4\rho(\rho - 1)u \quad \text{at } \zeta = 0 \qquad (4.39)$$

and the initial condition remains explicitly solvable,

$$\partial u/\partial \zeta = \kappa u \quad \text{at } \tau = 0, \qquad (4.40)$$

where
$$\kappa = 2\rho+1-M_i/(C_n - 1)$$
$$= 2\rho - \Theta_i / (C_n - \Theta_i) .$$

Similarly, the condition of undisturbed initial water content at infinite depth is

$$u^{-1} \partial u / \partial \zeta \quad \text{as } \zeta \to \infty .$$

By transformation (4.37), we may freely rescale the Hopf-Cole potential u, as this has no effect on the observable Kirchhoff variable M. Therefore, without loss of generality, the conditions (4.39) and (4.40) may be written explicitly as

$$u = \exp\,[\rho(\rho+1)\tau\,] \quad \text{at } \zeta = 0 \tag{4.41}$$

and

$$u = \exp\,(\kappa\,\zeta\,) \quad \text{at } \tau = 0. \tag{4.42}$$

By the method of Laplace transforms [e.g. Carslaw and Jaeger, 1959], we obtain the solution

$$u = \exp\left(\kappa\zeta + \kappa^2\tau\right) + \exp\left(-\frac{\zeta^2}{\tau}\right)\left\{\begin{array}{l} erc\left(\dfrac{\zeta}{\sqrt{\tau}} + [\rho(\rho+1)\tau]^{1/2}\right) \\ \quad + erc\left(\dfrac{\zeta}{\sqrt{\tau}} - [\rho(\rho+1)\tau]^{1/2}\right) \\ \quad - erc\left(\dfrac{\zeta}{\sqrt{\tau}} + \dfrac{\kappa}{2}\sqrt{\tau}\right) \\ \quad + erc\left(\dfrac{\zeta}{\sqrt{\tau}} - \dfrac{\kappa}{2}\sqrt{\tau}\right) \end{array}\right\} \tag{4.43}$$

In order to apply the Hopf-Cole transformation, we also need the derivative

$$\frac{\partial u}{\partial \zeta} = \kappa \exp\left(\kappa\zeta + \frac{1}{4}\kappa^2\tau\right) + \exp\left(-\frac{\zeta^2}{\tau}\right)\left\{\begin{array}{l} [\rho(\rho+1)]^{1/2}\, erc\left(\dfrac{\zeta}{\sqrt{\tau}} + [\rho(\rho+1)\tau]^{1/2}\right) \\ \quad - [\rho(\rho+1)]^{1/2}\, erc\left(\dfrac{\zeta}{\sqrt{\tau}} - [\rho(\rho+1)\tau]^{1/2}\right) \\ \quad - \dfrac{\kappa}{2}\, erc\left(\dfrac{\zeta}{\sqrt{\tau}} + \dfrac{\kappa}{2}\sqrt{\tau}\right) \\ \quad + \dfrac{\kappa}{2}\, erc\left(\dfrac{\zeta}{\sqrt{\tau}} - \dfrac{\kappa}{2}\sqrt{\tau}\right) \end{array}\right\} \tag{4.44}$$

The Hopf-Cole and Kirchhoff transformations may be inverted to give

$$\Theta = C_n\left[1-\left(2\rho+1-u^{-1}\right)^{-1}\right] \quad . \tag{4.45}$$

The dimensionless depth is obtained by inverting the Storm transformation:

$$z_* = [C_n(C_n - 1)]^{-1/2} \int_0^Z M dZ$$

$$= [\rho(\rho+1)\tau + (2\rho+1)\zeta - \ln u]/\, C_n \; . \tag{4.46}$$

Equations (4.43)-(4.46) constitute an exact solution to the constant-flux boundary value problem. This is an exact parametric solution, based on parameter ζ that runs from 0 to ∞. At a given time t_{s*} and choice of parameter value ζ, we may exactly locate a point (z_*,Θ) on the water content profile. In practice, we rarely need to evaluate these points at depth greater than around 2.5 times the Green-Ampt wetting front. At this point, ζ is approximately

$$[C_n - 1 + 1.5\{C_n - \Theta i\}]\; r_*\, t_{s*} \,/\, (1-\Theta_i) \; .$$

We cannot eliminate the parameter ζ to obtain an explicit analytic solution in terms of familiar functions. However, since we have defined ζ to be 0 at z=0, we can explicitly evaluate the water content at the soil surface. This includes evaluating the time evolution of θ_o during rainfall, and calculating the time to incipient ponding. In Appendix IV we present an example (MATLAB5) program which can calculate a profile of wetting based on Equations (4.43-4.46). Figure 4.4 shows one solution for each of a range of nonlinearity parameters, C_n.

Time to Ponding

During an infiltration event, the time to incipient ponding t_p is the elapsed time when the pressure potential at the soil surface first becomes zero. An increase to positive pressure could then be physically maintained only by an overlying pond. In this analytically solvable model that does not allow for a tension-saturated zone, zero pressure potential is the unique potential at which $\theta = \theta_s$. Incipient ponding must occur at some time if for all times, r exceeds K_s, or equivalently r_* exceeds 1. In this case the practically relevant range of moisture content must include θ_s as the least upper bound. Then the relation between time to ponding and application rate follows after equating Θ to 1 in equation (4.45). The emerging equation turns out to be most conveniently expressed in terms of the dimensionless ponding time that is non-dimensionalized in a particular way:

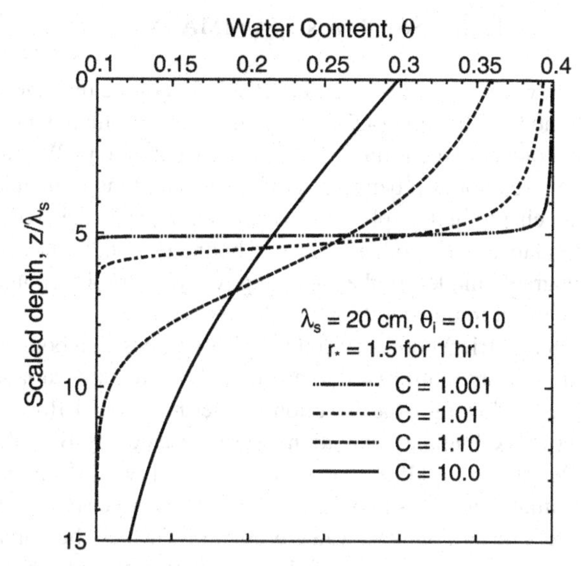

Figure 4.4. The solution of the system of equations (4.43-4.46) for flux infiltration for 4 values of C. The convection-dispersion equation behaves as expected in response to increased linearity with increased C values.

$$T_p = t_p / t_{grav} = t_p (K_s - K_n)^2 / S(\theta_n, \theta_S)^2 .$$

The full relationship that holds for any value of the nonlinearity parameter Cn and any value of initial water content, is

$$1 - \frac{2C_n}{r_*} =$$

$$\left\{1 - \frac{2C_n(C_n - 1)}{r_*}\frac{\Theta_i}{C_n - \Theta_i}\right\}\exp\left\{\begin{array}{c}\frac{C_n(C_n-1)\Theta_i^2}{(C_n-\Theta_i)^2}\frac{C_n(C_n-1)}{h}T_p \\ -\frac{C_n}{C_n-\Theta_i}\frac{C_n(C_n-1)}{h}r_* T_p\end{array}\right\}\cdot$$

$$erf\left\{-\frac{r_*}{2}\sqrt{\frac{T_p}{h}} + \frac{\Theta_i}{C_n - \Theta_i}\frac{C_n(C_n-1)}{\sqrt{h}}T_p^{1/2}\right\}$$

$$-\left(1 + \frac{4C_n(C_n-1)}{r_*}\right)^{1/2} erfc\left\{\frac{r_*}{2}\left[\frac{T_p}{h}\left(1 + \frac{4C_n(C_n-1)}{r_*}\right)\right]^{1/2}\right\} \quad (4.47)$$

We will discuss ponding time calculations more thoroughly in the next chapter, along with several approximations.

DISCUSSION AND SUMMARY

In this chapter, we have reviewed the exact solution of the realistic integrable unsaturated flow model, Equation (4.5), on a semi-infinite domain subject to uniform initial water content and constant-flux boundary conditions. We have examined the consequent relationship between rainfall rate and time to incipient ponding. The following chapter will explore this relation in regard to analytic approximations. It is natural to ask what other initial and boundary conditions can be treated within this integrable model. In the following, we will briefly summarize current knowledge relevant to this question.

When zero-flux (barrier) boundary conditions are imposed at both boundaries of a finite column, it is straightforward to evaluate the Kirchhoff variable at each boundary, ultimately allowing exact solution of the unsaturated flow equations. Solutions with zero flux boundary conditions were presented by Broadbridge and Rogers [1990], Warrick *et al.* [1990], and Sander *et al.* [1991]. In principle, any physically valid initial conditions may be imposed. The above sequence of transformations leads to a linear boundary value problem for which the Laplace transform solution may or may not be inverted. In the situation of moisture redistribution, the water content is not necessarily monotonic in time. Therefore the usual neglect of hysteresis effects in these models is a serious physical deficiency.

The problem of prescribed positive flux (irrigation) at the surface of a finite layer was also solved exactly but this required much more complicated mathematical techniques that are not easily generalized [Broadbridge *et al.*, 1988]. This led to the remarkable result that provided that the irrigation rate is more than double the saturated conductivity, saturation will occur at the supply surface before it occurs at the basement barrier, no matter how thin the layer. The practical limitation of this model is its failure to consider air compression effects at the impervious basement.

One may consider a constant negative flux boundary conditions to simulate the atmosphere-controlled phase of evaporation [Warrick *et al.*, 1990; Stewart and Broadbridge, 1999]. The solution of the analytical model in this case involves error functions of complex arguments. The main problem with the analytical model in the case is that we cannot rely too much on the model diffusivity in Equation (4.5) to give accurate predictions of water transport as evaporation drives water content towards zero. At water contents near zero (or θ_n) the soil water diffusivity may in fact be much less than the value $h(C)S^2/[(\theta_s - \theta_n)C + \theta_n]^2$ assumed by the analytic model.

Since we may, in principle, incorporate arbitrary initial conditions with constant flux boundary conditions, we may solve the analytic model not only under constant flux boundary conditions, but also under piecewise-constant boundary conditions [Warrick *et al.*, 1991].

At this time there is no generally accepted practical method for constructing exact solutions under arbitrarily prescribed variable flux boundary conditions.

Barry and Sander [1991] provided a practical approach to converting this problem to an integral equation that is more readily solved by iterative or numerical means. Broadbridge et al. [1996] demonstrated one of many ways to construct an exact solution to the integrable nonlinear Richards' equation from any solution to the linear diffusion equation. In this way, some useful variable-flux boundary value problems were solved. However, simple solutions to the linear diffusion equation often lead to overly complicated solutions of the nonlinear problem. It is not yet possible to prescribe the same variable flux boundary conditions on soils with different nonlinearity parameter.

As far as we are aware, attempts to incorporate other interesting boundary conditions, such as the constant concentration [θ_0] boundary condition, in exact solutions to realistic nonlinear models, not requiring numerical solutions of integro-differential equations at any stage, have not yet been successful. This subject is still very much alive, both from the theoretical and the practical point of view.

5

Absorption and Infiltration Relations and the Infiltrability-Depth Approximation

INTRODUCTION

In this chapter, the fundamentals of infiltration theory will be employed to derive analytic functions for infiltrability and cumulative infiltrated depth. The objective is to relate the properties of the dynamics of intake rates to soil hydraulic properties or parameters that directly represent them. The measurement of infiltration parameters that represent the soil hydraulic properties is discussed in a later chapter.

As indicated earlier, the focus here is on one-dimensional, usually vertical flow. While multidimensional versions of Richards' equation describe flows in a variety of cases, including flows from subsurface irrigation and upflow from a water table, we are interested primarily in obtaining tools for typical hydrologic purposes, i.e., flows of rainfall or surface water downward into soil. Fortunately, this process is generally one-dimensional. Some treatment of two- or three-dimensional flows will be important, however, considering the case of flow caused by (circular) infiltrometers or permeameters from which true one-dimensional flow fields cannot be created [Chapter 8]. As in Chapter 3, both absorption and infiltration conditions are examined, because of the importance that the generally simpler and more analytically tractable absorption case has in informing the approach to solution of the infiltration case.

In addition, the fundamental approximation concerning the stability of the flux concentration relation between the cases of ponded and flux infiltration is examined. On this, in effect, depends the quality of the important IDA principle [described by others as "time compression"]. Below, we will look at the quality of those approximations by solutions with realistic soil hydraulic characteristics, employing a precise numerical solution.

Infiltration Theory for Hydrologic Applications
Water Resources Monograph 15

Mass Balance Across the Intake Boundary

The mathematical basis of infiltration theory starts with an integral expression for continuity across the soil surface or intake boundary. This is in contrast with, but related to, the differential continuity expression that forms the basis of Richards' equation. Physically, the expression states that, during an infiltration event, the flux into the soil at the surface must equal the change in storage in the wetted area near the surface. We assume here for simplicity that the initial water content θ_i is uniform with depth. The surface flux balance may be expressed in differential form as

$$f = \frac{d}{dt}\int_0^{z_L} (\theta - \theta_i)dz \tag{5.1}$$

in which z_L is some depth beyond the depth of influence of the intake boundary ($z = 0$) condition. Calling $\boldsymbol{I}$ the infiltrated water depth in the wetted soil adjacent to the surface, an equivalent form is

$$\int_0^t f dt = \int_0^{z_L} (\theta - \theta_i)dz = I(t) - K_i t \tag{5.2}$$

The term $K_i t$ only applies to the infiltration case, and is the infiltrated depth due to the gravitational flux associated with the initial water content. Most often K_i is nearly zero, and is ignored, but it will be kept in the discussion here, since it becomes important in infiltration following a rainfall hiatus ("reinfiltration"), or any time the soil may have a relatively high water content, θ_i. For the absorption, or imbibition case, this term is zero in any case, since there is no gravitational flux.

In this chapter, as previously, we will deal with two types of intake boundary conditions creating inflow. Conditions at this boundary, where $z = 0$ (or $x = 0$), will be designated by the subscript "o." For the flux boundary condition, in which intake flux v_o is imposed, the value of θ_o will not be fixed, but will change with time. For a suddenly saturated boundary condition, θ_o will be imposed and will be a fixed upper limit for integration of a basic equation derived from Equation (5.2). The sorptivity, S, is a function of the surface water content and the initial water content. Since the surface condition may be less than saturation, the definition of sorptivity can be somewhat generalized from that given in Equation (3.12):

$$S(\theta_o) = \int_{\theta_i}^{\theta_o} \varphi \, d\theta \tag{5.3}$$

Note that S is implicitly a function of θ_i. When S is used without indications otherwise, it will be assumed that it refers to $S(\theta_s)$.

ABSORPTION

We begin with the simpler but yet important case of intake of water without the effect of gravity. The solution of the absorption equation (3.4) was presented briefly in Chapter 3, for constant head boundary conditions (3.6), and we briefly review this here. Philip [1957a] showed that the absorption case, described by Equation (3.2) or (3.4), for a given function for $D(\theta)$ (with certain reasonable limitations), has a similarity solution in which the water advance profiles $\theta(x,t)$ for all times scale as $t^{-1/2}$. This was expressed in Equation (3.8). Using the length x, for horizontal flow, rather than depth z, the application of Equation (5.2) to Equation (3.8) for fixed θ_0 boundary conditions yields Equation (3.10) or (3.12):

$$I = \left[\int_{\theta_i}^{\theta_o} \varphi \, d\theta\right] t^{1/2} = S(\theta_o) t^{1/2} \tag{3.12}$$

Now we may also replace $\boldsymbol{I}$ in Equation (3.12) with $\int_0^t f_c \, dt$, which is true by definition for this boundary condition, and differentiate both sides with respect to time to obtain, as above:

$$f_c = \frac{S}{2} t^{-1/2} \tag{3.14}$$

These two equations hold for any soil characteristics for which S may be evaluated, as discussed below. We may also eliminate time by combining Equations (3.12) and (3.14), to obtain a relation between the absorptibility, f_c, and the cumulative infiltrated depth, $\boldsymbol{I}$:

$$I = \frac{S^2}{2 f_c} \tag{5.4}$$

or:

$$f_c = \frac{S^2}{2 I} \tag{5.5}$$

These equations, it should be stressed, are found under the condition of a fixed pressure potential ψ (or a fixed water content θ_o) as the surface boundary condition. Equation (5.2) places no such condition of the intake boundary, with which we shall consider now the case of a surface condition with constant flux of value v_o. Under all conditions, the flow at any point in the wetting zone is described by Darcy's law, which for the gravity-free case may be written $v = -D d\theta/dx$. In addition, since we are now dealing with horizontal flow, we replace z, the length scale of Equation (5.2), with x. Rearranging Darcy's law into an expression for dx:

$$dx = -\frac{D}{v} d\theta \tag{5.6}$$

which is substituted into Equation (5.2), along with an appropriate change of integration limits, to obtain the *absorption integral*:

$$I = \int_{\theta_i}^{\theta} \frac{(\theta - \theta_i) D \, d\theta}{v(\theta, t)} \tag{5.7}$$

Note that v, as illustrated in Figure 5.1, refers to the flux at any point within the advancing wet zone. As θ is monotonically decreasing with z, $v(\theta)$ is the flux passing through the plane at point $\theta(z)$, and from conservation of mass is also the change in water content in the wetting region below level $\theta(z)$:

$$v(\theta, t) = \frac{\partial}{\partial t} \int_{\theta_i}^{\theta} z \, d\theta \tag{5.8}$$

An alternate method of deriving Equation (5.7) by integration of Richards' equation is presented in the Appendix.

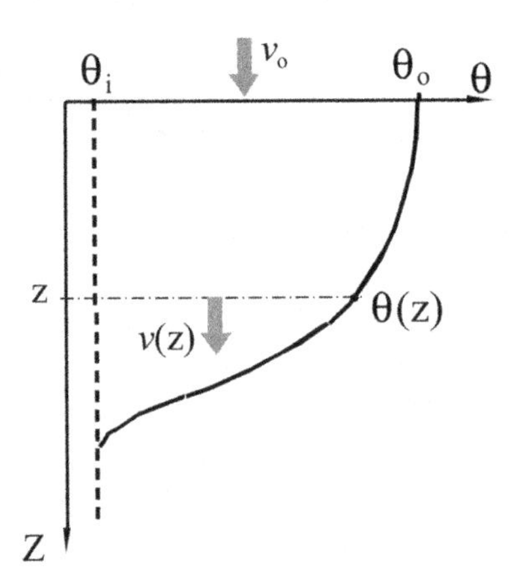

Figure 5.1. Definition diagram for the relation of θ to z during infiltration, and the flux at each value of θ, used for the flux-concentration relation.

The Flux-Concentration Relation

We now turn to examine how v varies with distance or with θ in the absorption case. To describe this relation, the behavior of flux within an advancing wetting wave has spawned an important variable called the *flux-concentration* relation [Philip, 1973], which is merely the flux at any point in the wetting $\theta(x)$ curve, scaled by that at the boundary (refer to Figure 5.1):

$$F = \frac{v}{v_o}\ (absorption)\ \ or\ F = \frac{v - K_i}{v_o - K_i}\ (infiltration) \tag{5.9}$$

in which v_0 represents the flux at the intake surface, and K_i is the initial (gravitational) flow, for the infiltration case only. This scaled flux is conveniently expressed as a function of scaled water content : $\Theta_i = (\theta - \theta_i)/(\theta_o - \theta_i)$, where θ_o is the surface water content, the upper bound for integration of Equation (5.7). This conditionally scaled water content is based on a lower limit of θ_i rather than θ_r, and is therefore distinguished from Θ_e of Equation (2.5).

With the definition of Equation (5.9), then Equation (5.7) may be rewritten, in general, as

$$I = \int_{\theta_i}^{\theta} \frac{(\theta - \theta_i) D\, d\theta}{v_o F(\theta, t)} \tag{5.10}$$

As Philip [1973] pointed out, because of the similarity of the absorption case solution, F may also be expressed in terms of the Bolzmann similarity variable:

$$F(\Theta_i) = \frac{\int_{\theta_i}^{\theta} \varphi\, d\theta}{\int_{\theta_i}^{\theta_o} \varphi\, d\theta} \tag{5.11}$$

For delta function diffusivity (introduced in Chapter 3), the function $F(\Theta_i)$ [Philip, 1973] is simply $F = \Theta_i$, but in fact for this special soil characteristic there are no values of v for θ less than θ_0. For a variety of other more realistic soil properties the flux-concentration relation is remarkably close to the relation $F = \Theta_i$. More will be said about these functions later, examples of which are illustrated in Figure 5.2.

For fixed θ_o boundary conditions, the upper limit for integration in Equation (5.7) is θ_o, and flux v is a variable in time and θ (or z). For this boundary condition the boundary value of v_o is the absorptibility $f_c(t)$. For an imposed surface flux, such as a rainfall rate r; the boundary value $v_o = f = r$. If, for a simple constant flux $v_o = r$ case, the assumption of $F = \Theta_i$ is made, Equation (5.7) immediately reduces to the simpler form:

$$I = rt = \frac{(\theta_o - \theta_i)}{r} \int_{\theta_i}^{\theta} D \, d\theta \tag{5.12}$$

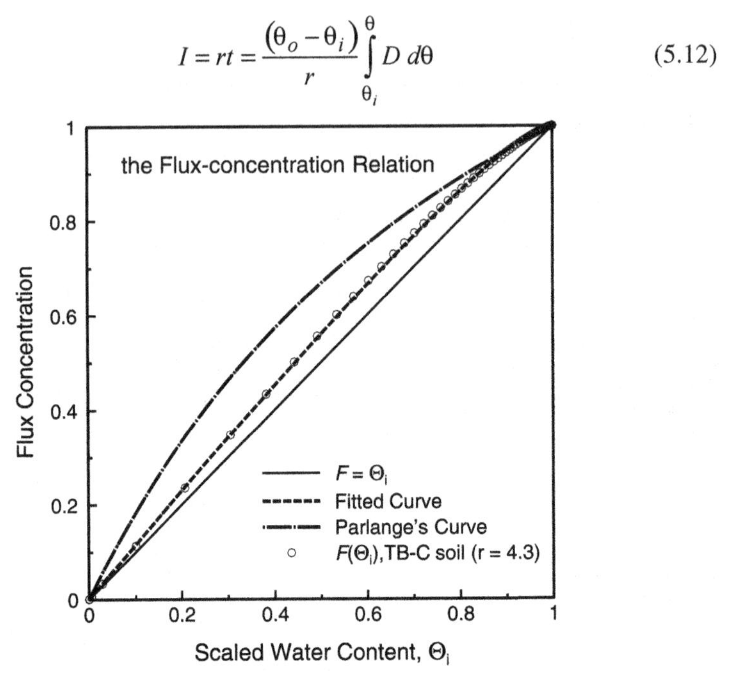

Figure 5.2. The flux-concentration relation, F(Θ), does not vary extensively as the soil properties change, and goes between (0,0) and (1,1), above the line F = Θ.

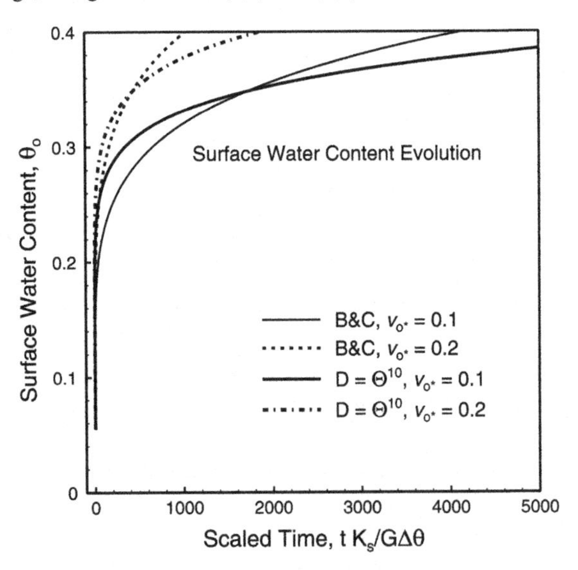

Figure 5.3. The evolution of surface water content as described by Equation (5.12), assuming F = Θ, depends on the soil hydraulic properties and the relative surface influx rate, $v_o{}^* = v_o/K_s$.

Given integratable expressions for D, this absorption equation may be solved to describe or approximate the evolution of the surface water content with time, as a function of *r* (or *f*). This will be illustrated below. The time evolution of θ_o thus determined for two typical soil characteristics is illustrated in Figure 5.3.

For fixed $\psi = 0$ or "constant concentration" boundary conditions, Equations (5.5) and (5.10) also produce an important relationship between diffusivity and sorptivity (both defined in Chapter 3, above). From these two equations one obtains

$$\frac{S^2}{2} = \int_{\theta_i}^{\theta_s} \frac{(\theta - \theta_i) D \, d\theta}{F} \tag{5.13}$$

The Capillary Length Scale

Equation (5.12) serves to help define a fundamental soil infiltration parameter, the *capillary length scale,* which depends on the soil hydraulic characteristics and the initial condition. This important measure was introduced in Chapter 4, for the case where initial water content was essentially zero, or $\theta_i = \theta_r$, and was termed λ_s. It was discussed in detail by Philip [1985] and White and Sully [1987], and was proposed as early as 1964 by Bouwer [1964]. Here we generalize this concept slightly and use the symbol G(•) to represent this value that may be dependent on θ_i or ψ_i. This integral soil parameter may be expressed in several equivalent forms. In most general terms,

$$G(\psi_i) = \frac{1}{(K_s - K_i)} \int_{\psi_i}^{0} K(\psi) d\psi \tag{5.14}$$

which for initially dry soils (K_i being negligible) becomes

$$G = \lambda_s = \frac{1}{K_s} \int_{-\infty}^{0} K(\psi) d\psi \tag{5.15a}$$

or in terms of the relative conductivity [Equation (2.16)],

$$G = \int_{-\infty}^{0} k_r(\psi) d\psi \tag{5.15b}$$

With the definition of D(θ) G may also be expressed as:

$$G(\theta_i) = \frac{1}{(K_s - K_i)} \int_{\theta_i}^{\theta_s} D \, d\theta \tag{5.16}$$

As pointed out by others [White and Sully, 1987], G or λ_s is effectively the k_r -weighted mean value of ψ.

In the constant flux boundary case of Equation (5.12), the surface water content, θ_o, increases steadily until the time when $\theta_o = \theta_s$. Now the surface water content can no longer increase in response to the inflow. At this time the boundary condition must change from one of constant flux to a ponded or constant θ_o condition. With the definition of Equation (5.16), and the step-function based $F = \Theta_i$ assumption, at the time when θ_o reaches θ_s, Equation (5.12) reduces to

$$I_p = \frac{(K_s - K_i)(\theta_s - \theta_i)G}{r} \tag{5.17}$$

Thus just after ponding, since the boundary condition of Equation (5.17) is a constant $\theta_o = \theta_s$, it is comparable with Equation (5.4). A similar expression also results from integrating the absorption integral Equation (5.7) for the constant θ_o boundary condition, wherein v_o is the time variable absorptibility f_c:

$$I = \frac{(K_s - K_i)(\theta_s - \theta_i)G}{f_c} \tag{5.18}$$

Either expression, (under the condition $F = \Theta_i$), yields the important relation of G (or λ_s) to S, comparing either equation to Equation (5.4) or (5.5):

$$G(\theta_s - \theta_i) = \frac{S^2}{2(K_s - K_i)} \tag{5.19}$$

For other (not delta-function) diffusivity (or F) functions, White and Sully [1987] suggested replacing the factor ½ in Equation (5.19) with a parameter, b, which varies from 0.5 to approximately 0.8:

$$G(\theta_s - \theta_i) = \frac{bS^2}{(K_s - K_i)} \tag{5.20}$$

The coefficient b may be formally defined [White and Sully, 1987] as follows, from the relations given above (Equations 5.10 and 5.12):

$$b = \frac{(\theta_s - \theta_i)\int_{\theta_i}^{\theta_o} D\,d\theta}{2\int_{\theta_i}^{\theta_o} \frac{(\theta - \theta_i)D\,d\theta}{F}} \tag{5.21}$$

Note that the upper boundary for integration in this expression is the fixed surface condition, but not necessarily saturation, such that a value of S (and b) can be determined for any fixed upper bound condition.

A small correction can be made to $F(\Theta_i)$ in determining the correct integral capillary value G for use in Equation (5.15) or (5.16). We express $F(\Theta_i)$ using a correction function δ_F:

$$F(\Theta_i) = \Theta_i\,\delta_F(\Theta_i) = \frac{\theta-\theta_i}{\theta_s-\theta_i}\delta_F(\Theta_i);\quad \delta_F \geq 1 \tag{5.22}$$

with $\delta_F(1) = 1.0$, and $\delta_F = 1.0$ for the delta-function case.. Note that δ_F can be treated as a function of θ, within appropriate limits, as well as Θ_i. The expression for G is then more accurately written [starting from Equation (5.13)] :

$$G = \frac{1}{(K_s-K_i)}\int_{\theta_i}^{\theta_s}\frac{D\,d\theta}{\delta_F(\theta)} \tag{5.23a}$$

or:

$$G = \frac{1}{K_s-K_i}\int_{-\infty}^{0}\frac{K\,d\psi}{\delta_F(\theta(\psi))} \tag{5.23b}$$

The integral effect of the correction δ_F can also be related to the ratio of b (Equation 5.21) to 0.5.

Most mathematical forms for $k_r(\psi)$ do not lend themselves to direct integration for G as indicated by Equation (5.15b) or (5.23b), including relations (2.19) and (2.21). However, the simple relation of Brooks and Corey, Equation (2.16), can be integrated to indicate limits to the value of G given the soil parameter ψ_B. For that expression of $k_r(\psi)$, integration obtains

$$G = \int_{-\infty}^{\psi_B}\left(\frac{\psi_B}{\psi}\right)^{\eta} d\psi + \psi_B = \frac{\eta}{\eta-1}\psi_B \tag{5.24}$$

subject to modification by a non-uniform value of δ_F. Since the curvature parameter c in the more realistic relation (2.21) serves to reduce this integral value, the value of G will in almost all cases vary within the limits ψ_B and 2 ψ_B.

The Infiltrability-Depth Approximation: IDA

It is important to note that the equality of the relations of Equations (5.4) and (5.17) is not in the time domain, although they can be converted into expressions in time. The implied result in Equation (5.17), in comparison with Equation (5.18), is that for the flux boundary condition $v_o = r$, boundary-limited absorption begins, and the boundary condition changes to a fixed θ_o case, when the infiltrated amount I_p is equal to that which occurs at the value of absorptibility = r for the case where the boundary condition is 'ponded' from the start. This is the fundamental value in using the relation $f(\boldsymbol{I})$ or $\boldsymbol{I}(f)$ rather than time [f(t) or $\boldsymbol{I}$(t)]. In the constant flux case (Equation 5.17), the ponding time t_p occurs at $\boldsymbol{I}_p = r{\cdot}t_p$, and infiltrated amounts and infiltrability f_c after that should be described by the relationship in Equation (5.18), provided only that the integrals in Equation (5.7) evaluate equally for the D(θ) function and the F functions of the two conditions. In the next chapter it will be demonstrated that the flux need not be constant prior to ponding. Thus the relation f_c ($\boldsymbol{I}$), rather than f_c(t), is valuable for estimating infiltration from a variable rainfall pattern.

This principle is fundamental to hydrologic applications of infiltration theory, and here we will refer to it as the infiltrability-depth approximation: *IDA*. Others have termed this the 'time compression approximation' (e.g. Sivapalan and Milly, [1989]), after the empirical method of Reeves and Miller [1975]. That is felt to be less appropriate terminology. Reeves and Miller [1975] did not look at $f_c(\boldsymbol{I})$ directly, but retained time as a common parameter in a computational method. Time is not being compressed, rather the surrogate independent variable, $\boldsymbol{I}$, is used rather than time. This change of independent variable is far more robust for dealing with both the flux and ponded boundary conditions and the transition from the former to the latter. The word approximation is used because, except for certain cases, $F(\Theta_i)$ is not exactly equal under the two boundary conditions.

Sivapalan and Milly [1989] have proposed that validity criteria for the *IDA* is equality and time invariance of $F(\Theta)$ functions for the two boundary conditions. In fact, *IDA* validity, not to mention its utility, should be assured for slightly less restrictive conditions. The approximation is exact for the delta-function case [Philip, 1973], as indicated already, where also $F = \Theta$. The approximation does not hold for a so-called "linear" or constant D soil, for which case the F

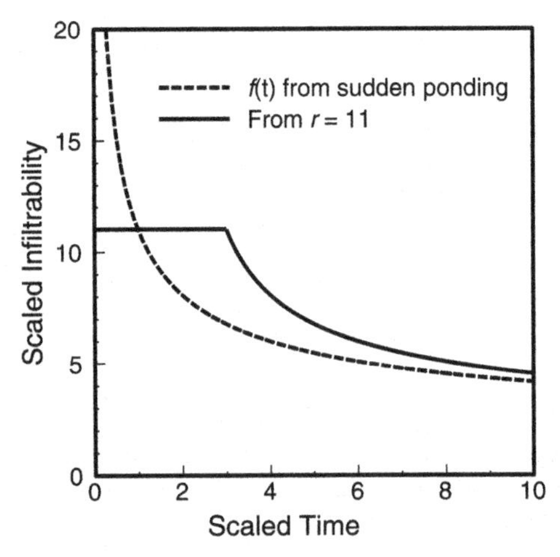

Figure 5.4. When infiltrability is plotted as a function of time, ponding during rainfall occurs at times which do not match the infiltrability relation from the ponded boundary condition.

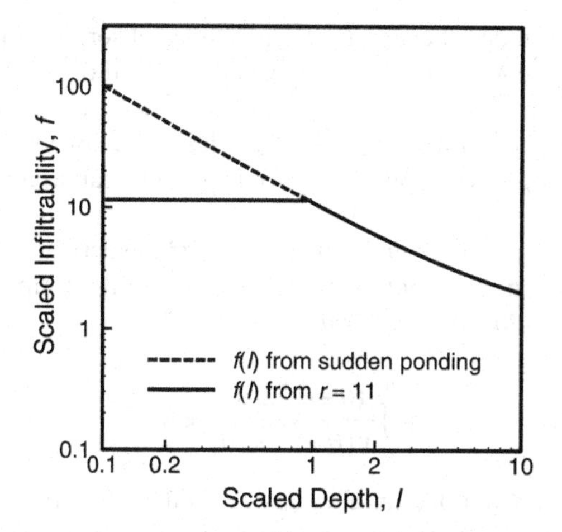

Figure 5.5. When infiltrabilities from rainfall and from flooding upper boundary conditions are compared as functions of infiltrated depth, I, rather than time, as in Figure 5.4, a unity of function is found for all portions of the curves where $I > I_p$.

functions for the two different boundary conditions can be obtained with Laplace solution methods [Philip, 1973]. Additionally, as time approaches ∞, all infiltration F functions approach linearity, as Philip[1973] showed. Philip [1973] also pointed out that $F(\Theta)$ should be independent of time for constant θ_o conditions for all cases where diffusivity is a simple power function: $D \propto \theta^m$. However, time independence is not to assure that F is similar for the two different boundary conditions. Since the infiltration relation depends on the integral in Equation (5.23), the *IDA* holds sufficiently for hydrologic purposes as long as G is essentially equal at $f_c = r(t_p)$. The success of the *IDA* for the power function D case, and the relatively small G variation for realistic soil characteristics, is demonstrated experimentally below.

Conversion of the infiltration relations for both boundary conditions into the time domain will describe the time to ponding at flux rate r in comparison to the $f_c(t)$ curve., as illustrated in Figures 5.4 and 5.5. It should be noted here that the relations above only described the gravity free case: for absorption, or for infiltration in early times when gravitational effects can be neglected. Now we turn our attention to infiltration: absorption plus gravity.

INFILTRATION

Much of what has been presented above for the absorption case serves as an introduction to the infiltration case. The utility of the IDA is the same for infiltration conditions. For infiltration, the boundary volume balance equation includes the flux due to higher values of θ_i when appropriate, and the expression for Darcy's law includes the gravity term. Under absorption, ponding -- the change of boundary condition when θ_o reaches θ_s -- will occur at some time for any positive value of flux *r*. For infiltration, as we will see, ponding cannot in theory occur for imposed fluxes equal to K_s or less, since due to the gravitational potential, a homogeneous soil with sufficient depth can accept water at that rate indefinitely.

Following the same arguments which preceded the derivation of the absorption integral, Equation (5.7), but using Darcy's law in the form Equation (2.3), one can derive the *infiltration integral*:

$$I = \int_{\theta_i}^{\theta_o} \frac{(\theta - \theta_i) D \, d\theta}{v(\theta, t) - K(\theta)} + K_i t \tag{5.25}$$

It may be remarked again that an alternate method of deriving Equation (5.25) [see Haverkamp, *et al.*, 1990] is to reformulate and integrate Richards' equation [Philip, 1969], and is reproduced in the Appendix. As was true for Equation (5.7), the infiltration integral is general, and can apply to both ponded and flux-type boundary conditions; *i.e.* θ_o fixed and v_o varying with time, or v_o constant and an evolving θ_o.

Time Series Solution of Philip

Following from the time series solution for $z(\theta,t)$ given in Equation (3.20), a time series solution for infiltration from a ponded (or constant θ) upper boundary was presented by Philip [1957a, 1969]. Just as was done in Equation (3.9) for absorption, an expression for the cumulative infiltration depth *I* may be obtained by integrating the time series solution for z over the wetted region, resulting in the following series expression:

$$I(t) - K_i t = \int_0^\infty (\theta - \theta_i) dz = S t^{1/2} + A_2 t + A_3 t^{3/2} + A_4 t^2 + \ldots.. \tag{5.26}$$

As indicated earlier, this solution is quite commonly used in truncated form, with K_i usually neglected. Only the first two terms of the series are retained, and K_s is used for A_2. As can be seen by Equation (3.21), however, A_2 is not K_s in the series; e.g. the truncated expression has A_2 approximately $2K_s/3$ when a delta-function soil is used. Other soil characteristic assumptions allowing mathematical tractability produce other values for A_2 [Philip, 1987]. As Philip [1969] pointed out, the truncated expression is asymptotically correct in the short time (the first term), but correct in the long term only if the coefficient A_2 is made equal to K_s. However, it is a considerably biased estimate of infiltration flux rate through intermediate times, as demonstrated below. For this reason Philip [1987] described various approaches for correcting the truncated expression. The modified truncated series expression is usually referred to as the Philip infiltration expression:

$$I = St^{1/2} + K_s t \tag{5.27}$$

For clarity in the following discussions, the symbol $\boldsymbol{I'}$ will be used for the infiltrated depth due to surface flux exclusive of the term $<K_i t>$. This term can affect computations in some cases, but is rarely significant. Note that in Equation (5.27) the left side represents $\boldsymbol{I}$ rather than $\boldsymbol{I'}$, since the K_i terms cancel from both sides. This equation is differentiated to obtain

$$f_c = K_s + \frac{S}{2} t^{-1/2} \tag{5.28}$$

These two equations are directly comparable to Eqs. (5.4) and (5.5). Time may easily be eliminated between Equations (5.27) and (5.28) to find, with a little algebra, the corresponding 'Philip' expression for $f_c(\boldsymbol{I})$:

$$f_c = \frac{K_s \sqrt{IK_s + S^2/4}}{\sqrt{IK_s + S^2/4} - S/2} \tag{5.29}$$

Approximate Integration of the Infiltration Integral

The additional term in the denominator of the infiltration integral, Equation (5.25), prevents the kind of integration for simple but reasonable D functions that are possible for absorption through Equation (5.12), above. There are, however, at least three approaches for integrating this equation that are fundamental, and produce hydrologically valuable infiltration functions. For this purpose it is useful to restate the infiltration integral (Equation 5.9) into the following form, by introducing the flux-concentration relation, $F(\theta,t)$, corresponding to the absorption integral, Equation (5.10):

$$I = \int_{\theta_i}^{\theta_o} \frac{(\theta - \theta_i) D \, d\theta}{F(\theta,t) v_o - K(\theta)} + K_i t \tag{5.30}$$

As indicated above, the term $K_i t$ is included for completeness. While in most cases the value of K_i is several orders of magnitude too small to be considered, there will be cases treated below, particularly the case of reinfiltration after a storm hiatus, in which the equation will need to account for significant values of K_i.

In many cases of approximation for solving this equation, the flux-concentration F may be assumed to be equal to Θ_i. This is justified to the extent that soils are very nonlinear in $K(\psi)$ and may act like a delta-function D, or piston flow soil. As stated above, in the strict delta-function case, there are no points for the F relation other than θ_i and θ_o. The simplest $F = \Theta_i$ is also approached at long times for all soils. This can be shown in several ways. One demonstration is by integrating both sides of Equation (3.23), which describes the profile at large time, with respect to θ between θ_i and $\theta_o - \varepsilon$, using integration by parts under the general principle

$$\int_L^U \left[\int_x^U V(x)\, dx \right] dx = \int_L^U (x - L) V\, dx \tag{5.31}$$

Use of this tool for $v_o = u_o(\theta_o - \theta_i)$ constant (a traveling wave) obtains the infiltration integral and thus shows that $F = \Theta_i$ assymptotically at large time. Parlange (1975) has often assumed $F = 2\Theta_i/(\Theta_i\text{-}1)$ (Figure 5.2), a quite reasonable general approximation. We will demonstrate F functions for a variety of soils below.

Delta Function Approximate Integration As in the absorption case, one tractable assumption is that the diffusivity is closely represented by a Dirac delta function of Equation (3.29). This presumes that the soil diffusivity only contributes significantly to the integral (5.25) in a small region near saturation. In consequence, the term $(\theta - \theta_i)$ only contributes at this value of θ, and the values of $v(\theta,t)$ and $K(\theta)$ are single-valued as well. With the delta-function assumption for D, one may formally integrate the expression with $K(\theta) = K_s$, *i.e.*, as a delta function. Since $K(\psi)$ [and $v(z)$] is a step function, it may be more meaningful to integrate the expression over ψ rather than θ. In any case, under this assumption for D, the integral (5.30) describes a moving 'piston' wave from the surface to $z = \boldsymbol{I}/(\theta_s - \theta_i)$, with a step change in K at the front from K_s to K_i. The result is precisely the well known Green and Ampt [1911] infiltration model (including the term for K_i):

$$I' = \frac{S^2}{2(f_c - K_s)} = \frac{G(\theta_s - \theta_i)(K_s - K_i)}{f_c - K_s} \tag{5.32}$$

As indicated above, since for this delta-function assumption the flux-concentration relation is $F(\Theta_i) = \Theta_i$, Equation (5.32) strictly obeys the *IDA* and applies to both ponded and flux surface conditions. Letting $f_c = r_p$, the rainrate at which ponding is achieved, this expression describes the infiltrated depth $\boldsymbol{I}_p$ when ponding is reached for an input rate $r_p > K_s$:

$$I'_p = \frac{G\Delta\theta_{si}(K_s - K_i)}{r_p - K_s} \tag{5.33}$$

For brevity here and following, the term $\Delta\theta_{si}$ is used to represent initial saturation deficit (θ_s - θ_i). As noted above, I' represents I - $K_i t$. It should be noted that theory does not require that the value of r be constant from time t = 0 up to ponding, but simply describes the relation between the cumulated depth of infiltration and the occurrence of ponding at a given rate r. There is in fact a realistic practical limit on how much change in r can take place immediately prior to ponding without severely affecting the predictive accuracy of this and similar equations. The wetting zone immediately adjacent to the surface must adjust to changes in r(t), but can adjust quite rapidly in most cases.

Equation (5.32) can of course be inverted to provide a relation of $f_c(I)$:

$$f_c = \frac{(K_s - K_i)(G\Delta\theta_{si} + I')}{I'} + K_i \tag{5.34}$$

Exponential K(ψ) Approximate Integration This method [Talsma and Parlange, 1972] employs the Gardner exponential K(ψ) relation given in Equation (3.30) to represent the nonlinearity of this relation, and takes advantage of the fact that, under this particular relation,

$$D d\theta = \frac{1}{\alpha} dK = G\, dK \tag{5.35}$$

Use of the Gardner relation is largely a mathematical convenience to enable integration, while the resulting expression has applicability for a wide range of other soil hydraulic functions which display a rapidly dropping value of K with decreasing ψ. The integration also implicitly assumes that $F = \Theta$ describes accurately enough the variation of v - K_i within the wetting profile. Under these assumptions, Equation (5.25) may be written with change in the variable of integration as follows:

$$I' = G\Delta\theta_{si} \int_{K_i}^{K_o} \frac{dK}{v_o - K} \tag{5.36}$$

At the time of ponding under a flux surface condition, or for a ponded surface, the upper limit of integration is K_s, and the integration result is [Parlange and Smith, 1976]:

$$I' = G\Delta\theta_{si} \ln\left(\frac{f_c - K_i}{f_c - K_s}\right) \tag{5.37}$$

Smith and Parlange(1978) derived a time implicit expression for f_c(t) specifically to treat times greater than t_p, with $I_p = rt_p$ calculated by Equation (5.37). That derivation employed assumptions similar to those made above but followed different methods than for Equation (5.37). Their expression, with the addition of terms for K_i, and using $K'_s = K_s - K_i$, is:

$$K_s'(t-t_p) = G\Delta\theta_{si}\left[\ln\left(\frac{(r_p-K_s)(f-K_i)}{(f_c-K_s)(r_p-K_i)}\right) - \frac{K_s'}{f_c-K_i} + \frac{K_s'}{r_p-K_i}\right] \tag{5.38}$$

One may show, as follows, that Equation (5.38) is equivalent to Equation (5.37), following the *IDA*. Referring to Figure 5.6, in which the time is plotted as a function of infiltration flux f, in keeping with the implicit nature of Equation (5.38), the infiltrated depth $\boldsymbol{I} - \boldsymbol{I}_p$ from time t_p to t, is indicated by the shaded portion of the graph in this figure, and may be found by integrating along the f axis as follows:

$$I_1' - I_p' = (f_1 - K_i)(t_1 - t_p) + \int_{f_1}^{r_p}\left[t(f) - t_p\right]df \tag{5.39}$$

Substituting for $(t - t_p)$ from Equation (5.38) into Equation (5.39), and integrating, the expression finally reduces to exactly represent an equivalent form of Equation (5.37):

$$I' - I_p' = G\Delta\theta_{si}\left[\ln\left(\frac{f_c-K_i}{f_c-K_s}\right) - \ln\left(\frac{r_p-K_i}{r_p-K_s}\right)\right] \tag{5.40}$$

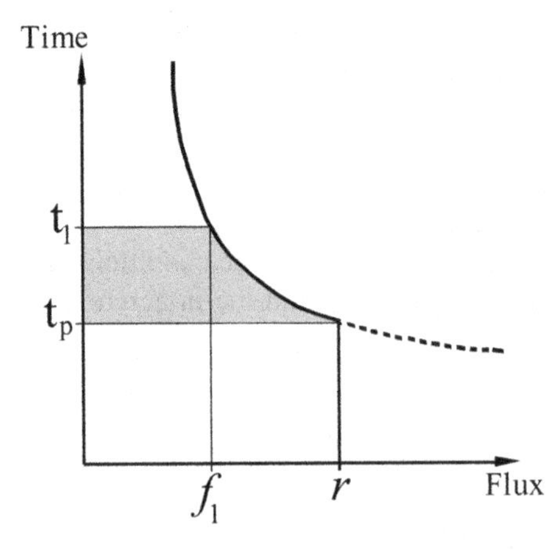

Figure 5.6. The shaded portion on this diagram, where time is plotted as a function of flux, represents the depth of water $I_p' - I'$ calculated in Equations (5.38) and (5.39).

Equation (5.37) also may be inverted to express infiltrability as a function of infiltrated depth:

$$f_c = \frac{K_s \exp\left(\frac{I'}{G\Delta\theta_{si}}\right) - K_i}{\exp\left(\frac{I'}{G\Delta\theta_{si}}\right) - 1} \tag{5.41}$$

In a similar manner, Smith and Parlange [1978]derived the time-implicit Green- Ampt equation using t_p predicted by Equation (5.33). Time implicit and explicit expressions will be treated in more detail in the following chapter.

Combination of Assumptions: the Three-Parameter Model The delta-function [Green-Ampt] model assumes, as shown in Figure 3.4b, that there is a region of ψ near 0 with constant K. The alternate [Smith-Parlange] model, Equation (5.37), assumes that $K(\psi)$, rather than being nearly constant at values of ψ near 0, drops very fast as ψ decreases from $\psi = 0$. Although these two assumptions are useful, and widely used, the characterization of soil properties presumed by them are mathematical caricatures of actual soil hydraulic functions. In fact, the original Brooks and Corey (B-C) function for $K(\psi)$ is much more a combination of these two assumptions: as ψ decreases from 0, K is constant up to the value of ψ_B, after which it drops very rapidly (although not exponentially, but as a power function).

The two infiltration models derived above are illustrated in Figure 5.7. This plot is for scaled variables, and we shall use scaled variables extensively in the following chapter, where the normalizing values are defined. This figure illustrates the differences in the two functions. Notice that the two formulas exhibit relatively large differences in the values of infiltrated depth, ***I***, for a given value of f_c at the intermediate ***I*** region as scaled f_c approaches the final value of 1. This asymptote is also demonstrated in Figure 5.7.

Parlange *et al.* [1982] have demonstrated a method by which an infiltrability model is produced that encompasses both the delta-function and the exponential $K(\psi)$ model behavior. To accomplish this, a third parameter of range 0 to 1 is found which effectively interpolates between the two K functions as extreme conditions. Starting with the basic infiltration integral, Equation (5.30), a function for K is judiciously selected with the additional parameter, here termed γ, as an interpolating weighting factor. First, we define the surrogate variable $U(\theta)$:

$$U(\vartheta) = \int_{\theta_i}^{\vartheta} \frac{(\vartheta - \theta_o)D(\vartheta)d\vartheta}{F(\vartheta)} \tag{5.42}$$

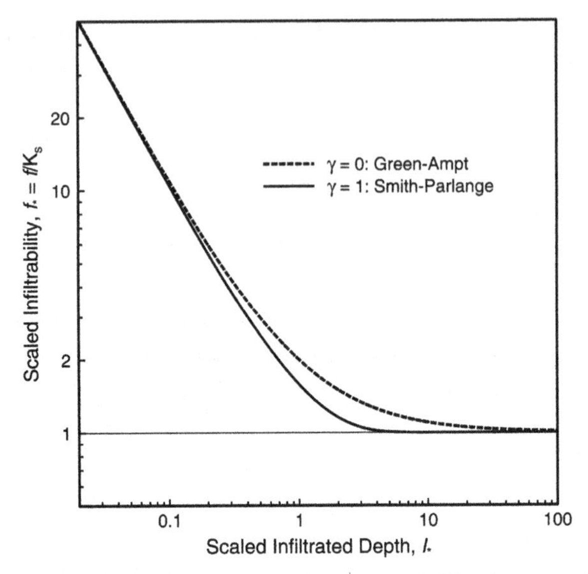

Figure 5.7. The third parameter, γ, in the 3-parameter infiltration equation covers the behavior expected by all real soils, which should exhibit infiltration curves that fall within the limiting curves shown here.

Note that $U(\theta_o) = S^2/2$, [from Equation (5.13)] and $U(\theta_i) = 0$. Using this, an expression for K is adopted with weighting parameter γ which encompasses both assumptions as extremes:

$$K(\theta) - K_i = (K_s - K_i)F(\Theta_i)\left[1 - \frac{2\gamma}{S^2}U(\theta)\right] \tag{5.43}$$

Parlange et al. [1982] explain this expression to some degree by showing that it may be integrated between θ_i and θ_s, using integration by parts as in Equation (5.31), to produce a somewhat clearer expression:

$$\gamma(\theta_s - \theta_i)(K_s - K_i) \approx \int_{\theta_i}^{\theta_s}(K_s - K)d\theta \tag{5.44}$$

Clearly, $\gamma = 0$ implies $K = K_s$. Parlange *et al.* [1982] argue that for γ to approach 1, K must approach a delta-function, i.e., be near K_i for most of the range of θ, and increase steeply γ as θ approaches θ_s. Based on Equation (5.44), however, no real soil would γ have as large as 1.

First we treat the ponded case where the upper limit of integration is θ_s. When Equation (5.30) is recast in terms of U as defined in Equation (5.42), and Equation (5.43) is used for K, we have

$$I' = \int_{U(\theta_i)}^{U(\theta_s)} \frac{dU}{F(\Theta)\left(f_c - K_i - (K_s - K_i)\left[1 - \frac{2\gamma}{S^2} U\right]\right)} \quad (5.45)$$

This expression, with the approximation $F = \Theta_i$, may be straightforwardly integrated to produce the 3-parameter infiltration equation [Parlange *et al.*, 1982]:

$$I' = \frac{S^2}{2\gamma(K_s - K_i)} \ln\left[\frac{f_c - K_i - (K_s - K_i)(1-\gamma)}{f_c - K_s}\right] \quad (5.46)$$

which may also be written, using Equation (5.19) :

$$I' = \frac{\Delta\theta_{si} G}{\gamma} \ln\left[1 + \frac{\gamma(K_s - K_i)}{f_c - K_s}\right] \quad (5.47)$$

By inspection, when γ approaches 1, Equation (5.47) approaches the Smith-Parlange expression of Equation (5.37). As γ approaches 0, one can also show by series expansion of the form $ln(1 + \gamma a)$ that the equation approaches the Green-Ampt expression, Equation (5.33), as expected. Demonstrations in Parlange *et al.* [1982], and other experimental and numerical results, indicate that a value of γ on the order of 0.8 to 0.85 is commonly a best fit for normal soils.

Describing Surface Water Content Evolution Under Rainfall

The Smith-Parlange Approximation. Equation (5.36) may be rewritten to approximate the evolution of surface water content θ_o using an assumed $K(\theta)$ relation. As an example, the common relation $K = K_s\theta_e^{\varepsilon}$, (Equation 2.17), not at all inconsistent with Equation (5.35), may be used with a rain of intensity r, to integrate Equation (5.36) to a value of $K(\theta_o)$ less than K_s. For simplicity one may change the integration variable from K to $k_r = K/K_s$., and define $r_* = r/K_s$ and $r' = r - K_i$. Then Equation (5.36) becomes

$$r't = G\ \Delta\theta_{si} \int_{k_{ri}}^{k_r} \frac{dk_r}{r_* - k_r} \quad (5.48)$$

This expression is integrated to obtain, analogous to Equation (5.37);

$$\frac{r't}{G\Delta\theta_{si}} = \ln\left(\frac{r_*}{r_* - k_r(\Theta_e)}\right) \quad (5.49)$$

To be more accurate, for values of r_* less than 1.0, the value of G should be treated as a function of the asymptotic θ_o, for which the upper limit of integration of Equation (5.23) changes. For such values of $r_* < 1$, ultimate $\theta_o = \theta_u$ can be found immediately from the soil characteristic relation between k_r and θ, such as Equation (2.17). If $G(\theta_u)$ is however taken as a constant, Equation (5.49) may be solved for k_r and then, by substitution from Equation (2.17) for θ_o as a function of k_r, one may obtain an expression for $\theta_o(t)$:

$$\theta_o(t) = \theta_r + (\theta_s - \theta_r)\left[\frac{r_*\left(\exp(r't/G(\theta_u)\Delta\theta_{ui}) - 1\right) + k_{ri}}{\exp(r't/G(\theta_u)\Delta\theta_{ui})}\right]^{1/\varepsilon} \tag{5.50}$$

As expected, the term in brackets will reach a value of 1.0 at some time t_p, when ponding occurs. Also, for t = 0 it correctly has $\theta = \theta_i$. Note that for all r_* less than or equal to 1 the water content will not ever reach saturation, which is correct yet quite unlike the corresponding expression for absorption as illustrated in Figure 5.3, above. The function is plotted in Figure 5.8 for a few values of r_*. While this relation may not be exact during constant flux wetting (being based on approximate integration, using fixed G and linear *F*), it is correct in its limits for all positive values of *r*, finding $\theta_o = \theta_s$ at ponding time, and having the proper asymptotic θ_o for all $r < K_s$. Figure 5.9 demonstrates, in normalized form, the relatively good simulation of the surface water content by this expression. Here a nor-

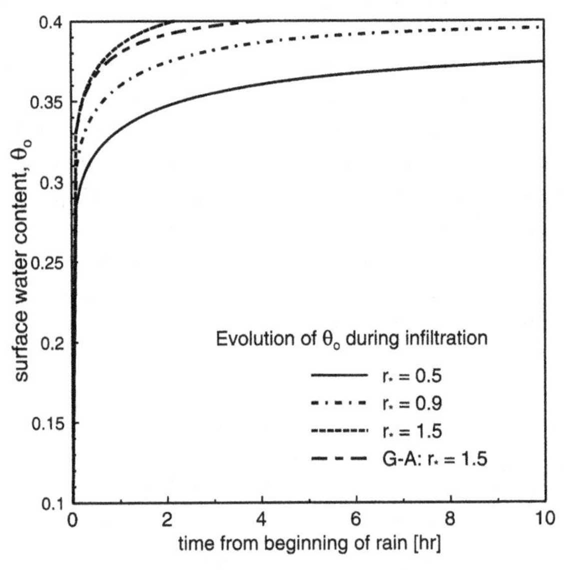

Figure 5.8. The infiltration integral yields Equation (5.50) which describes the evolution of surface water content as shown here for a range of relative rainfall rates, $r_* = r/K_s$.

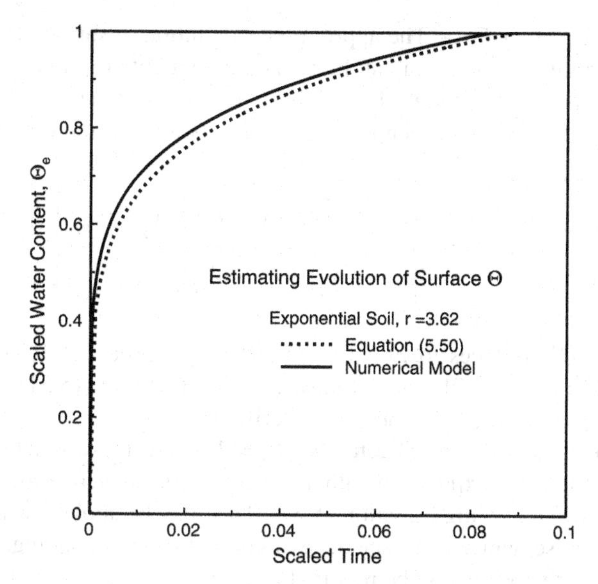

Figure 5.9. The approximate infiltration integral predicts the evolution of surface water content relatively accurately compared with the full solution of Richards' equation, as shown here for one example.

malized flux rate of 3.6 for an example case is compared with a numerical solution of Richards' equation solved for the same conditions.

The delta-function/Green-Ampt Case. The above result for the evolution of surface water content was obtained formally by an integration of the approximated infiltration integral to upper integration limits, in this case k_r, less than saturated. The step-function case does not allow such intermediate integral limits, since formally K is either 0 or K_s. However, the form of Equation (5.49) suggests that a comparable function may be found for the Green-Ampt delta-function D soil. In that case, the integration is not formally performed to an intermediate value of k_r or θ, but the delta function expression, when treated as Equation (5.49) above, produces a comparable function for $\theta_o(t)$ which again has correct limiting behavior. Following the procedure used above, we replace K_s in Equation (5.33) by $k_r(\Theta_e(t))$ and $\boldsymbol{I}$ by rt, then solve the expression for k_r. By inverting this $k_r[\theta_o(t)]$ relation, we obtain a function comparable to Equation (5.50), which describes reasonable behavior of the surface water content evolution for either $r > K_s$ or $r < K_s$, given the G-A assumptions:

$$\theta_o(t) = \theta_r + \left(\theta_s - \theta_r\right)\left[\frac{r_*(r't) + G\Delta\theta_{si}k_{ri}}{G\Delta\theta_{si} + r't}\right]^{1/\varepsilon} \tag{5.51}$$

One example of this G-A based $\theta_o(t)$ function is shown in Figure 5.8.

The 3-Parameter Case. The approximate evolution of surface water content for this integrating assumption can be found, comparable to Equation (5.50), in the same manner: integrating Equation (5.45) up to arbitrary θ with $K(\theta)$ expressed in terms of $k_r(\theta)$, then finding $\theta(k_r)$ as before, which obtains:

$$\theta_o(t) = \theta_r + \left(\theta_s - \theta_r\right)\left[\frac{r_* \left(\exp\left[\gamma\, r' t / G\Delta\theta_{si}\right] - 1\right) + \gamma k_{ri}}{\exp\left[\gamma\, r' t / G\Delta\theta_{si}\right] - 1 + \gamma}\right]^{1/\varepsilon} \quad (5.52)$$

As expected, this reduces to Equation (5.50) as γ approaches 1, and becomes essentially Equation (5.51) as γ vanishes. Thus Equation (5.51), which was obtained conceptually, is asymptotically verified.

For a sequence of rainfall rates, $r(j)$, j = 1,2,....n, these functions may be applied stepwise. An expression (albeit somewhat messy) for $d\theta_o/dt$ may be obtained by differentiating Equation (5.52). This may be applied sequentially to each rainfall pulse, which can trace the evolution of θ_o until ponding, if it occurs. Runge-Kutta integration may be required.

The Broadbridge-White Soil As indicated in Chapter 4, Equations (4.43)-(4.46) with $\zeta = 0$ describe the evolution of surface water content under rainfall rate *r*. The solution of these equations for a Broadbridge-White soil with C = 1.01 to obtain the evolution of surface water content is shown in Figure 5.8. While the early time values of θ_o vary from those of Equations (5.50) and (5.51), the B-W and S-P examples actually reach θ_s (ponding) at nearly the same time: about 4.3 h. The early time variations simply reflects the difference in the shape of the $k_r(\theta)$ relations for the two soils.

Treatment of Surface Water Depth

The effect of additional small surface head h_s can be approximated by substituting $(h_s + G)$ for G in the infiltrability expressions. An additional method for this purpose was recently introduced by Haverkamp *et al.* [1990]. They reasoned that there would, under positive surface pressures, be a small zone near the soil surface that would be saturated and need to be described by a modified expression. The Haverkamp *et al.* [1990] model is in effect a method to account for a tension saturated zone such as would occur if real soils behaved like the B-C functions, with a singular point at ψ_B, and with saturation for all greater values of ψ. Arguing that all soils exhibit some tension saturated depth, they added a term to Equation (5.47) explicitly for this:

$$I' = \frac{(h_o + h_s)(K_s - K_i)\Delta\theta_{si}}{f_c - K_s} + \frac{\Delta\theta_{si} G}{\gamma}\ln\left[1 + \frac{\gamma(K_s - K_i)}{f_c - K_s}\right] \quad (5.53)$$

The additional term models flow in the presumed tension-saturated zone of depth z_S using a Green-Ampt type piston flow model with $h_s + h_o$ in place of G. The resulting expression (5.53), however, cannot be inverted to obtain an explicit $f_c(\boldsymbol{I})$ relation as was done in Equations (5.34) and (5.41)

Without the assumption of a tension-saturated zone, treatment of surface depths are conceptually straightforward, replacing the capillary drive parameter G with $G + h_s$. This is effective where surface depths may be significant, such infiltration from surface irrigation.

STABILITY OF FLUX-CONCENTRATION RELATION.

Integration of the infiltration integral to produce the analytic infiltration equations discussed above tentatively adopted the approximation $F(\Theta_i) = \Theta_i$. Further, the infiltrability-depth approximation, *IDA*, in strictest terms presumes that $F(\Theta_i)$ is equal under ponded and flux boundary conditions. It was clearly recognized that this is not exact for soils other than the delta-function D soil. Therefore before continuing with applications of these infiltrability relations following the *IDA*, it is useful to evaluate the degree of variation in $F(\Theta_i)$ between flux and fixed θ_0 (or ψ_o) boundary conditions, and the degree to which this may affect accuracy of the *IDA* in finding an infiltrability pattern during a rainfall.

While $F(\Theta_i)$ is time stable for ponded (fixed pressure) absorption, and thus is expected to be relatively invariant for infiltrability at short times, the extent of time dependence under infiltration will need to be assessed. The extent to which basic assumptions are violated, and moreover the effect of such violations, will have importance for the confidence we may have in the use of the *IDA* in calculating infiltration for applications and hydrologic design. It is more important to understand just what effect non-linear and time-varying characteristics the $F(\Theta_i)$ function has on the infiltration behavior of soils with hydraulic properties that may be encountered in nature.

Below we examine the linearity and stability in the $F(\Theta_i)$ relation for 3 different soil types, using both ponded and flux boundary conditions. From a considerable experience of measured soil properties, we know the van Genuchten or transitional Brooks-Corey (TB-C) soil hydraulic functions are good models of most soil properties. By using these and two others with considerably different (and perhaps extreme) mathematical shape, however, one may gain some insight into the sensitivity of the *IDA* concept to nonlinearity and variability of the $F(\Theta_i)$ function.

The method used to examine the $F(\Theta_i)$ is a relatively precise numerical solution of Richards' Equation which allows selection of soil hydraulic properties as well as upper boundary conditions. The essential features of this numerical method are described in the appendix. It has been verified against a quasi-analytic solution for the absorption case [suggested by Philip, 1974], and the analytic solution of Broadbridge and White [1988] [Smith, 1990], Appendix 3. The numerical results, as seen below in a few cases, are subject to highest sensitivity at the bottom of the advancing θ front, where all the hydraulic characteristics are changing very rapidly with depth. Fortunately, it is this part of the solution which has the least, indeed almost negligible, effect on the value of the infiltration or absorption integral, since relative conductivities and fluxes are very small in this region. The upper parts of the flux-concentration F curves are smooth and stable and are well supported by the quasi-analytic solution when comparable.

Soil Types Evaluated

The delta-function soil may be considered one extreme nonlinear soil characteristic assumption, but it is in many cases a reasonably close approximation, and it is also one for which the *IDA* holds exactly. Conversely, an unrealistic constant D soil is one for which it is known that the *IDA* does not hold [Philip, 1973]. However, the $F(\Theta_i)$ relation for a constant D soil is obtainable by analytic solution, for both ponded and flux conditions [Philip, (1973)]. These $F(\Theta_i)$ relations are illustrated in Figure 5.10. For normal soil hydraulic relations the transitional Brooks-Corey functions given in Equations (2.20) and (2.21) will be used. Another different soil characteristic relation is the exponential relation for $k_r(\psi)$, Equation (3.30). This was used above to help derive the Smith-Parlange model. The soil with this $k_r(\psi)$ relation, along with the normal retention relation of Equation (2.20), will be called the *quasi-exponential* soil type.

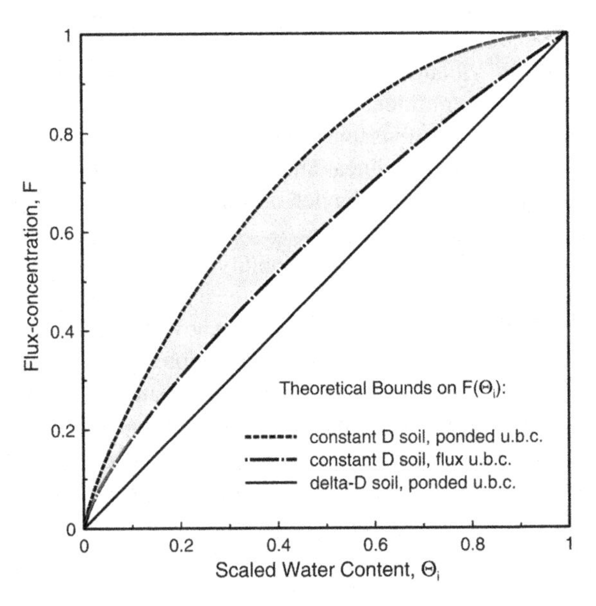

Figure 5.10. The flux-concentration relation for real soils should lie between the limits shown here as the dashed and solid lines. Extreme violations of the IDA assumption by the constant D soil are represented by changes of F(θ) in the shaded portion of this graph.

A third type that we use to study the behavior of the flux-concentration [FC] relation has both the exponential $k_r(\psi)$ [Equation (3.30)] and an exponential retention function. To obtain this retention function we assume that the relative conductivity is a power function of the normalized relative saturation, Θ_e, as for the Brooks-Corey model of Equation (2.17). Then we have:

$$k_r = e^{\alpha\psi} = \Theta_e^{\varepsilon}$$

which then provides the retention relation:

$$\Theta_e = e^{\frac{\alpha}{\varepsilon}\psi} \tag{5.54}$$

This soil type will be called an *exponential* soil type. From the definition of $D(\theta)$, one may show that for this soil the diffusivity is also a power function:

$$D(\Theta_e) = \frac{K_s \varepsilon}{\alpha} \Theta_e^{\varepsilon - 1} \tag{5.55}$$

The characteristics of the quasi-exponential and the exponential soils are illustrated in Figures 5.11 and 5.12. While the two soils may not appear dramatically different in these graphs, the mathematical relation between relative conductivity and saturation is different and cannot be made very similar by parameter manipulation. Table 5.1 lists the parameters of the soils used in the FC experiments below.

F(Θ_i) Under Absorption

Because of the similarity solution presented in Chapter 3, Equation (3.10), absorption under a constant head (or θ) boundary should have a flux- concentration relation that is stable with time. This is observed in the numerical solution

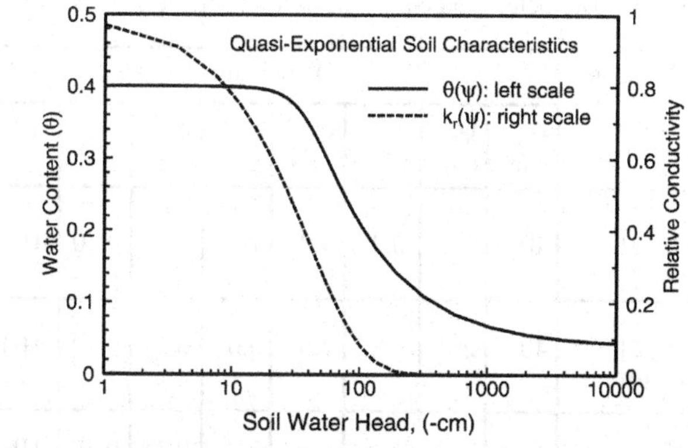

Figure 5.11. The hydraulic characteristics of the quasi-exponential soil used in evaluation of the IDA assumption. These curves represent Equations (2.16) and (3.30) used together.

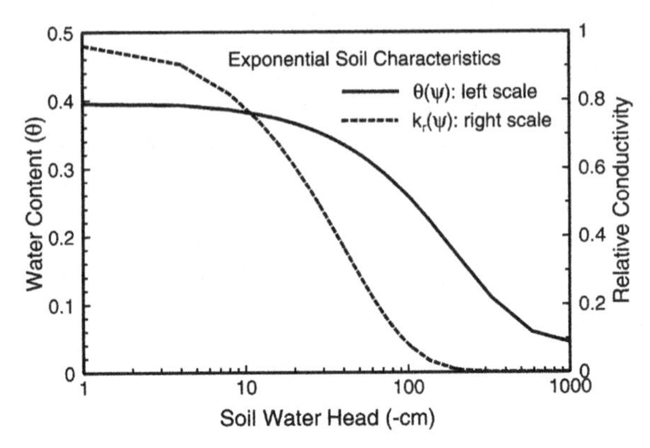

Figure 5.12. Graphical display of the hydraulic characteristics of the fully exponential soil used in evaluation of the IDA assumption. These curves represent Equations (5.54) and (3.30).

results which are presented here, and this is shown by example in Figure 5.13. In addition, an iterative method for closely calculating the flux-concentration relation for this case was presented by Philip and Knight [1974], whose results (the "quasi-analytic solution") are compared with the numerical model results in this figure.

Philip [1973] indicates that the upper curve and the straight line in Figure 5.10 represent the limits within which soil $F(\Theta_i)$ relations should lie. Recalling the absorption model developed above, the assumption inherent in the simple IDA model requires that the cumulative infiltrated (absorped) depth, $\boldsymbol{I}$, with a ponded boundary condition $[\theta_o(t>0) = \theta_s]$, at the point where $f = r$, should be equal to the cumulative depth at the point of ponding, $\boldsymbol{I}_p$, for flux boundary $v_o = r$. The subsidiary requirement to $\boldsymbol{I}(r) = \boldsymbol{I}_p$ is that $F(\Theta_i)$ be essentially equal for the two instances [Sivapalan and Milly, 1989]. For the cases of the exponential and quasi-

Table 5.1. Descriptive parameters and equations used in soils tested for F(θ) properties

Soil	Equations	Parameters Used								
		ψ_B cm	ψ_a cm	λ_a	c	ε	α cm^{-1}	θ_s	θ_r	θ_i
TB-C	[2.16] [2.17]	40	2	0.2	4.0	13.		0.40	0.04	0.19
Quasi-Exponential	[2.16] [3.30]	40	2.	0.8	4.0	5.0	0.25	0.40	0.04	0.067
Exponential	[5.53] [3.30]	-	-	-	-	-	0.25	0.40	0.04	0.042

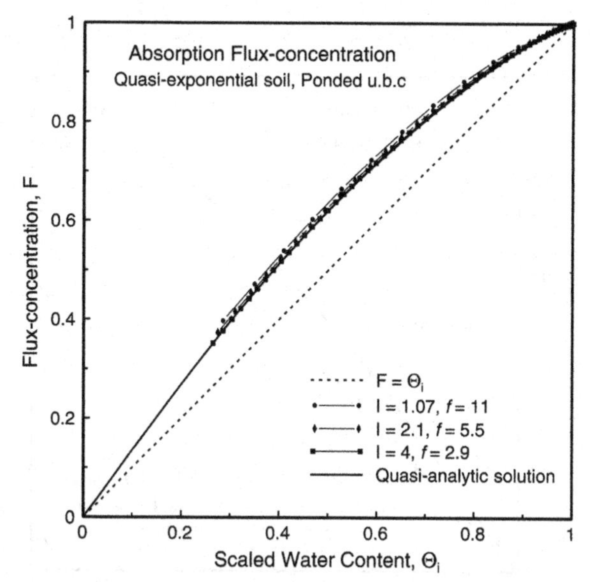

Figure 5.13. The flux-concentration relation remains relatively steady with time during imbibition from a ponded boundary condition.

exponential soil, the two $F(\Theta_i)$ functions are indeed distinctly different, as illustrated in Figure 5.14a and b. Figure 5.15 is the corresponding case for the more common TB-C soil, and also shows a difference between $F(\Theta_i)$ for ponded and flux upper boundary conditions (u.b.c.), albeit a much smaller difference. In each soil, it is curious how close to linear the $F(\Theta_i)$ function remains during flux absorption.

Figure 5.15 illustrates, however, that in the flux boundary case the soil wetting profile adjusts, immediately after ponding is attained, so that within minutes the flux-concentration curve becomes essentially equal to that for the initially ponded boundary. This rapid adjustment is reflected in the absorptability $f_c(\boldsymbol{I})$ relation for all the soil hydraulic functions, shown in Figures 5.16 through 5.18. It is in the faithfulness of this absorption (and infiltration) model relationship that the utility of assuming the IDA is best judged. As indicated above, an $F(\Theta_i)$ relation not equal to $F = \Theta_i$ can be reflected in the model relationship by simply finding a more appropriate G parameter value using Equation (5.20) or (5.23).

In fact, inspection of the $f_c(\boldsymbol{I})$ relations, shows that either option will not create significant errors in hydrologic applications, especially in comparison with typical field uncertainties always present. Furthermore, the case of the more normal TB-C soil exhibits such little difference between the ponding value $\boldsymbol{I}_p$ for the two $F(\Theta_i)$ relations (or the alternate estimates of G) that the IDA assumption is sufficiently accurate for normal calculation purposes.

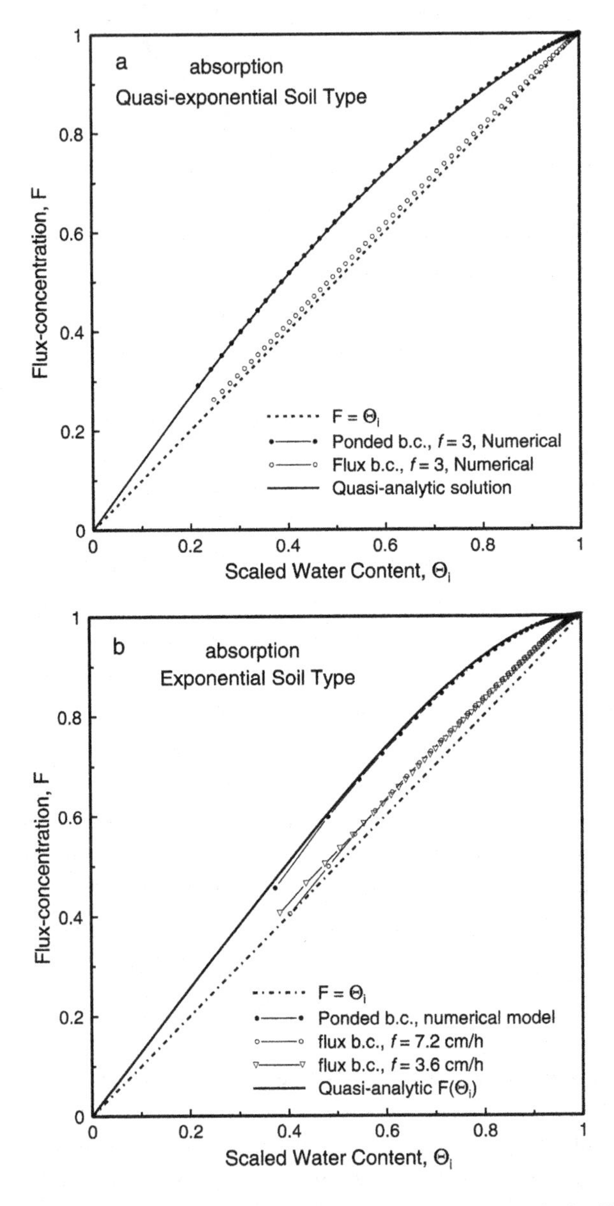

Figure 5.14. Both exponential (b) and quasi-exponential (a) soils exhibit significant changes in the flux-concentration relations between the ponded and flux boundary conditions during imbibition.

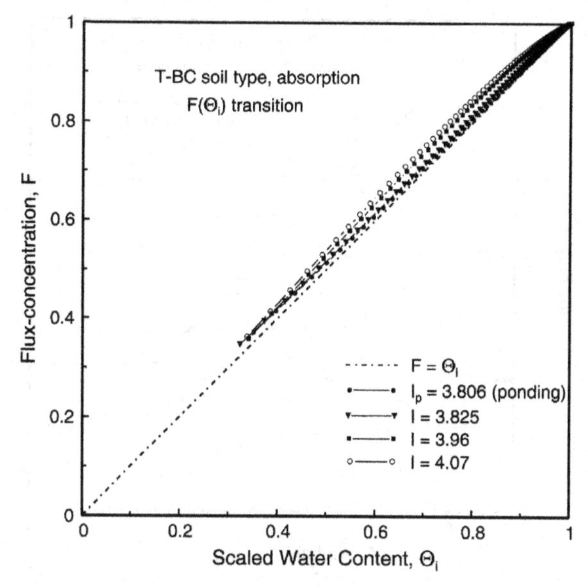

Figure 5.15. The flux concentration relations for soil characteristics having properties more like real soils, such as the T-BC function, have far less differences between ponded and flux boundary conditions, as illustrated in this figure for the case just after ponding from a flux of 3 cm/h.

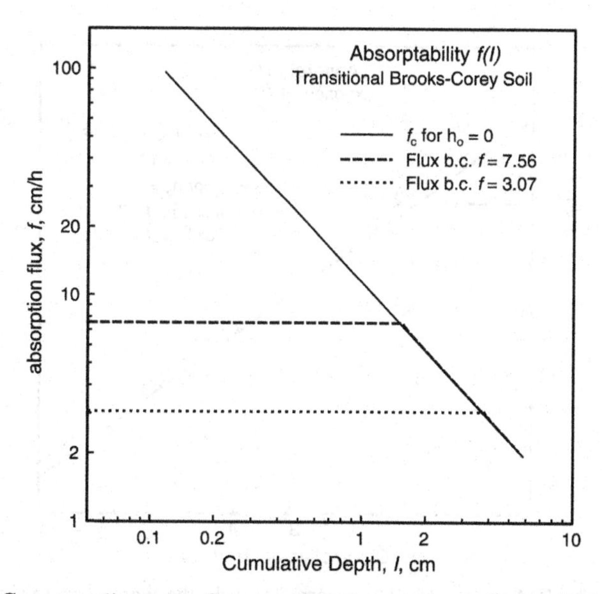

Figure 5.16. Corresponding to the flux-concentration relation in Figure 5.15, the absorptability relation for the transitional Brooks-Corey type soil exhibits near unity between the two boundary conditions, confirming the IDA method.

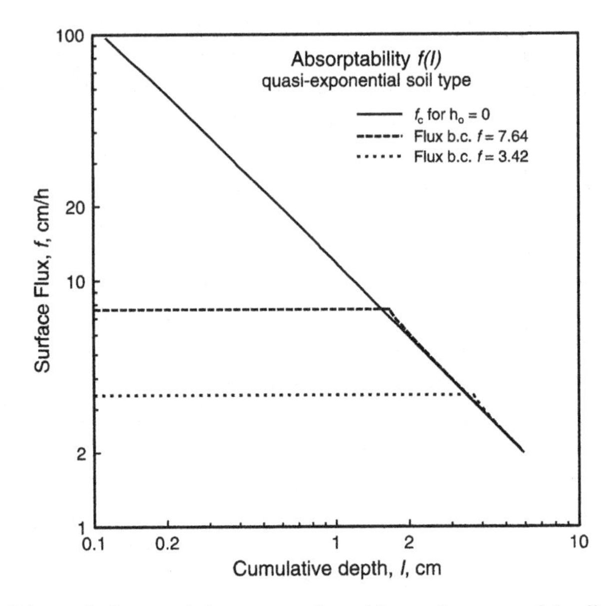

Figure 5.17. Other soil characteristic types, such as this quasi-exponential soil, do not justify the IDA as well as the more realistic T-BC soils (Figure 5.16), but are nevertheless close approximations to it, despite the differences in flux-concentration between the two boundary conditions.

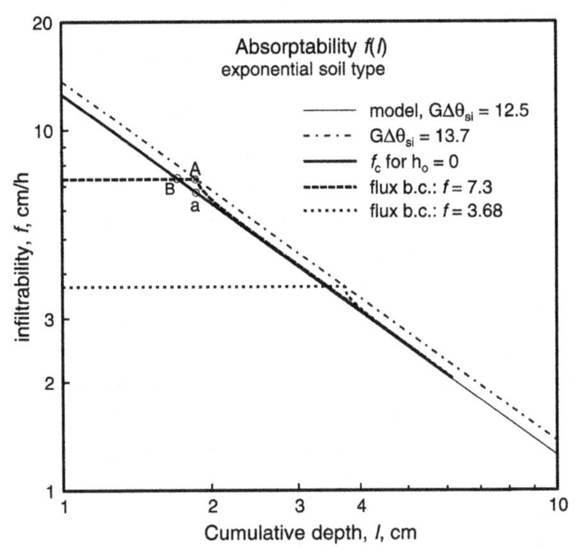

Figure 5.18. The IDA inaccuracy for absorptability in the worst case [exponential] soil type, shown here by the points near B-A-a, is still small in comparison to cumulative absorption depth, and can be treated in alternate ways discussed in the text.

$F(\Theta)$ Under Infiltration

The results above are essentially applicable to the infiltration case as well. The theoretical time stability of the $F(\Theta_i)$ relation under absorption does not follow, however, for infiltration. As indicated above, the $F(\Theta_i)$ curve will in all cases be slowly transient towards the line $F = \Theta_i$ at nearly infinite time. The amount of change in $F(\Theta_i)$ for the TB-C soil type over a reasonable time span is illustrated in Figure 5.19. It should be noted that most of the change occurs in the lower values of Θ_i and at the smaller values of $\boldsymbol{I}$. This part of the $F(\Theta_i)$ relation represents a tiny part of the actual $\theta(z)$ profile, but moreover the contributions to the infiltration integral, Equation (5.25), occur mainly in the upper values of Θ_i (and thus θ), so these changes do not result in notable errors in the *IDA* infiltrability calculations. Figure 5.20 illustrates the numerical solution for the $f(\boldsymbol{I})$ relation for two values of r, and the change in G due to shift in $F(\Theta_i)$ shown in this figure is as imperceptible as for the absorption case shown in Figure 5.16.

Implications of $F(\Theta)$ Observations

The foregoing results give a comprehensive (but not exhaustive) picture of the relative effect of boundary conditions, soil hydraulic properties, and time on the stability of $F(\Theta_i)$, and through the use of $F(\Theta_i)$ in Equation (5.23), the stabilty of G across boundary conditions changes. In terms of the linearity and stability of $F(\Theta_i)$, the two extreme soil properties are the delta function soil, where $F = \Theta_i$, and the constant D soil, for which $F(\Theta_i)$ may be analytically obtained. These two apparent limits were illustrated in Figure 5.10. This figure shows approximate curves from White *et al.* [1979] for F_p (constant head boundary) and F_f (constant flux boundary). The worst case, where D is constant, is quite far removed from the behavior of real soils, and should not concern us. It has nevertheless been studied in respect to the IDA validity [Philip, 1973; Liu *et al.*, 1998]. The use of $F_p(\Theta_i)$ for the flux case for this hypothetical soil underestimates the time to ponding by about 19%. More recent published results for more appropriately nonlinear D soils are found in [Parlange *et al.*, 2000] .

Briefly, we look at the apparent error involved for changes in $F(\Theta_i)$ for a realistic soil. The value of G found by use of Equation (5.23) using $F(\Theta_i)$ for the flux case may be termed G_f, and the corresponding G for the ponded case can be termed G_p. In light of the change in $F(\Theta_i)$ at $\boldsymbol{I}_p$, there are two strategies for treating the infiltrability model for the rainfall case:

A. One can estimate ponding using G_f, and make the shift to the use of G_p when ponding is attained (points A-a in Figure 5.18). The small error in f_c occurs only for a brief period just after ponding.

B. One can use the G_p to find a point of ponding which is earlier than strict theory, but is on the correct intersection of the infiltrability curve and the rainfall

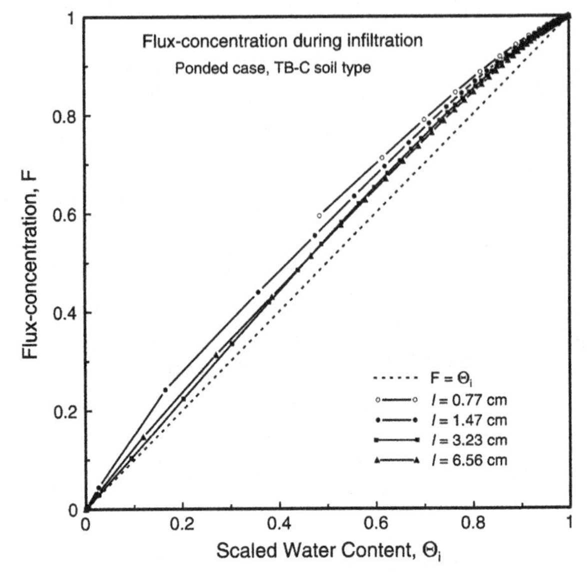

Figure 5.19. Flux-concentration during ponded infiltration into a T-BC soil type. Note that for this highly nonlinear soil the $F(\theta_i)$ relation changes minimally during increases in $\boldsymbol{I}$.

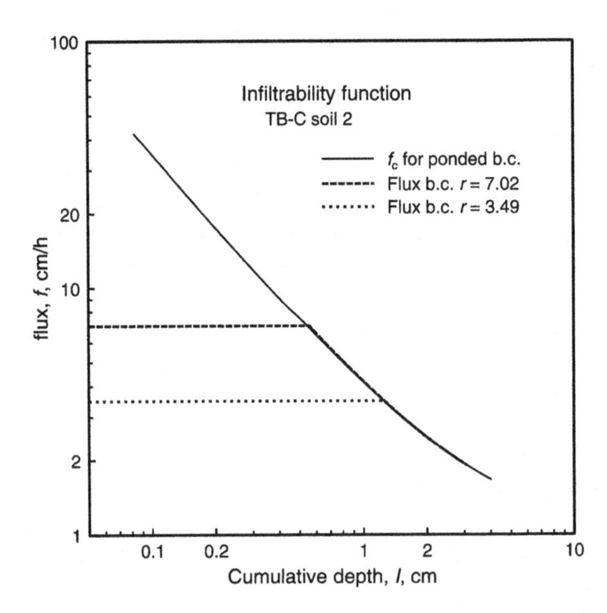

Figure 5.20. The infiltrability relation $f_c(I)$ shows very negligible effects of the changes in $F(\Theta_i)$ at the change in upper boundary conditions for the T-BC soil, in justification for the IDA method.

flux curve (point b in Figure 5.18). Only a small error, in comparison to the Richards' solution, occurs until the two f curves would rejoin.

Notable errors of this kind only occur for the assumption of soil properties that are generally quite unlike any of those in nature. The bias from approach B, indicating an earlier ponding than in theory, is conveniently in the direction of the effect of natural spatial variability, as discussed in Chapter 9. Another important observation, relative to this kind of potential error, is that there is a small error but no bias if G_p is used with the *IDA* method, but if G_f is used, a bias in estimated f_c does result. Fortunately, the G used in applications will not likely be obtained from use of a known $F(\Theta_i)$ or measured soil hydraulic relations, as in Equations (5.20) or (5.23), but from field measurements in which a constant ψ will normally be applied at the soil surface [Chapter 8]. This means that a measure of the value of G_p is likely to be the value obtained from field experiments, and this will produce less *IDA* estimation error for f_c after ponding.

More important than the worst case (and non-existent) linear soil case is the case of normal soils (Figures 5.16 and 5.20). The results of the foregoing survey have shown that for normal soils, changes in $F(\Theta_i)$ from flux conditions to ponded conditions are relatively slight and the effect is difficult to detect in the $f_c(\boldsymbol{I})$ relation. In any case, the robustness of the *IDA* in estimating both ponding occurrence and the evolution of infiltrability is more important than the small errors that may or may not be observed. This is especially true in light of the large uncertainty and spatial heterogeneity within which hydrologists must work in applying infiltration theory to real problems.

SUMMARY

This chapter has presented the derivation of the absorptibility equations, and three infiltrability equations from basic soil water dynamic equations: Darcy's law and the dynamic conservation of mass. The absorptibility equation is exact, and approximations or different forms for different soil hydraulic relations are not necessary. The concept of the soil flux-concentration relation, $F(\Theta,t)$, has been defined and used in the expression of the infiltrability integral equation. Two approximating assumptions on the relation of K to θ (or ψ) have been used to integrate the infiltration integral. The resulting expressions describe the evolution of water content under flux (rainfall) boundary conditions, and the time at which the boundary conditions must change, i.e., the ponding time.

The piston flow or Green-Ampt model, and the exponential K- derived Smith- Parlange equation can be thought of as representing two limiting assumptions about the way conductivity of the soil varies with water content. The 3-parameter equation which includes and interpolates between these two extremes is an optimal equation for allowing selection of an infiltrability formula to match a soil hydraulic behavior. Each of these approximate solutions to the infiltration integral also can be reformulated into an expression for the evolution of soil surface water content, by employing an appropriate relation for K(θ), and prediction of time to ponding for uniform rainfall rate.

The three infiltrability relations derived herein can be expressed in terms of any pair of the variables ***I***, f_c, and t. In the next chapter, all of these forms will be presented and compared in normalized form.

Since the approximate models are based on the *IDA* principle, implying stable [or at least equal $F(\Theta_i)$] for both upper boundary conditions, the *IDA* concept was evaluated by analysis of $F(\Theta_i)$ for a variety of soil characteristics to assure that significant errors should not be expected when using the *IDA* model for normal soils. The results of the Richards' solution comparison indicate that the flux-concentration relation does indeed vary slightly with time during flux infiltration, as expected. However, the variation is only slightly different between ponded and flux boundary conditions, even for the hypothetical exponential soils, and the strongly nonlinear nature of the underlying soil hydraulic characteristics also insures that small variation in the $F(\Theta_i)$ relation is relatively insignificant in effect. Apparently the *IDA* model is relatively quite accurate as a description of ponding and subsequent $f_c(\boldsymbol{I})$. The next chapter compares scaled forms of the equations with pairs of the variables *f*, ***I***, and t, and their application with the *IDA*.

6

Infiltrability Models: Comparisons and Application

INTRODUCTION

In the preceding chapter, several functions describing relations between the infiltration variables $\boldsymbol{I}$ and f_c were developed. Each analytic expression, deriving from different integrating assumptions, produces a relation of f_c , $\boldsymbol{I}$, and t, but not all of those forms were presented. In this chapter the other forms will be derived, and all are presented in normalized form. First the comparison of the various functions will be made in terms of simple ponding time for a uniform rainfall. In addition, we will explore the development of time-explicit forms of the relations [f_c(t)]. This form of an infiltrability equation is valuable in irrigation and other applications where $f = f_c$ during the time period of interest. There are a few time explicit forms f_c(t) that are quite accurate models of the time-implicit equations [t(f_c)], such as Equation (5.38). Finally, the application of the IDA equations to a rainfall of arbitrary rate pattern is discussed.

Scaling Parameters

All the infiltrability relationships presented up to now have common parameters: $G\Delta\theta_{si}$ (which expresses a parameter and a state variable, but these can be treated together) and K_s (and K_i). It is useful for comparison and for ease of use and manipulation to use normalized or scaled variables. The normalizing values used below are just the above physically-related parameters: $G\Delta\theta_{si}$ is the length scale or normalizing factor for infiltrated depth $\boldsymbol{I'}$. Thus the effect of initial water deficit, $\Delta\theta_{si}$, is included in the scaling in this manner. (K_s -K_i) is the normalizing scale for flux. In addition, we use a time scale related to one which Philip [1969, p.251] termed a characteristic time for infiltration, t_{grav}:

Infiltration Theory for Hydrologic Applications
Water Resources Monograph 15

$$t_{grav} \equiv \left(\frac{S}{K_s - K_i} \right)^2 \tag{6.1}$$

Here rather we use a characteristic time t_c based on the other normalizing variables:

$$t_c \equiv \frac{S^2}{2(K_s - K_i)^2} \equiv \frac{G\,\Delta\theta_{si}}{K_s - K_i} \tag{6.2}$$

so that $t_c = ½\, t_{grav}$. With these scaling factors, we define normalized values of f_c, $\boldsymbol{I}$, and t:

$$f_* = \frac{f_c - K_i}{K_s - K_i} \tag{6.3a}$$

$$I'_* = \frac{I'}{G\Delta\theta_{si}} = \frac{I - K_i t}{G\Delta\theta_{si}} \tag{6.3b}$$

$$t_* = \frac{t}{t_c} = \frac{t(K_s - K_i)}{G\Delta\theta_{si}} \tag{6.3c}$$

Note in Equation (6.3a) that for the scaled value the subscript c has been dropped for simplicity, but unless indicated otherwise, in the following discussions f_* represents the scaled infiltrability.

TIME OF PONDING

Under the appropriate range of rainfall rates, the end of the evolution of surface water content comes when, under boundary conditions (3.17), the surface water content or capillary potential can no longer increase. This is variously called the *time to incipient ponding*, or *time of ponding*, or *ponding time*, t_p. After this time the boundary condition must change to that of (3.16). Each of the relations derived in Chapter 5 with integrating approximations, as well as the exact integration of Broadbridge and White [1988] in Chapter 4, can be used to calculate the value of t_p. For the infiltration case and a uniform soil, ponding can only occur for $v_o = r > K_s$. For the absorption case, ponding or boundary saturation under conditions (3.6) will occur at some time under any positive value of v_o. The IDA provides a means to evaluate t_p for any r(t) pattern, as discussed below. Here we examine briefly the simple case of uniform r, which provides for a simple expression for t_p based on the $\boldsymbol{I}(f_c)$, and allows comparison with the analytic solution for the Broadbridge-White soil.

The Three Approximate Forms

The IDA provides that under uniform flux r, equations (5.32), (5.37) or (5.47), relating $\boldsymbol{I}$ to f_c, can be applied at ponding with $f_c = r$, and $\boldsymbol{I} = rt_p$. In that case, the ponding times are respectively, for the Green-Ampt or delta-function soils (recasting Equation 5.33) ,

$$t_p = \frac{G\Delta\theta_{si}(K_s - K_i)}{(r-K_s)(r-K_i)} \tag{6.4}$$

for the exponential K(ψ) Smith-Parlange case (Equation 5.37) ,

$$t_p = \frac{G\Delta\theta_{si}}{(r-K_i)} \ln\left(\frac{r-K_i}{r-K_s}\right) \tag{6.5}$$

and from the three-parameter general approximation, Equation (5.47):

$$t_p = \frac{G\Delta\theta_{si}}{\gamma(r-K_i)} \ln\left[1 + \frac{\gamma(K_s - K_i)}{(r-K_s)}\right] \tag{6.6}$$

Similar equations can be written for the respective absorption cases given above.

When scaling parameters given in Equation (6.3) are used, normalized forms of these expressions for time of ponding under uniform rainfall rate are formed:

Green and Ampt:

$$t_{p*} = \frac{1}{r_*(r_* - 1)} \tag{6.7}$$

Smith and Parlange:

$$t_{p*} = \frac{1}{r_*} \ln\left(\frac{r_*}{r_* - 1}\right) \tag{6.8}$$

and the three-parameter general approximation,

$$t_{p*} = \frac{1}{\gamma r_*} \ln\left(1 + \frac{\gamma}{r_* - 1}\right) \tag{6.9}$$

in which $r_* = (r - K_i)/(K_s - K_i)$. The comparison of these normalized functions is shown in Figure 6.1.

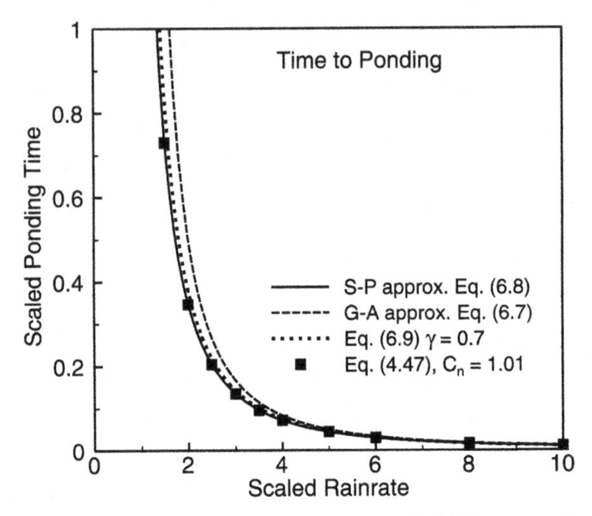

Figure 6.1. Several ponding time relations are compared. All here use the time normalization Equations (6.3), so that comparison is facilitated. Clearly, differences are negligible at larger rain rates.

The Broadbridge-White Soil

The exact integral available for the Broadbridge-White soil given in Chapter 4 will also determine a time to ponding, and gives us another way to evaluate the quality of the approximations in the approximate integral forms above. In the limit as $C_n \rightarrow 1$, the normalized ponding time relation (4.45) from Broadbridge and White[1988] reduces to

$$t_{p*} = (1 - \Theta_i)\frac{1}{r_*}\ln\left[\frac{r_*}{r_* - 1}\right] \qquad (6.10)$$

For flow models with delta function diffusivity, we have:

$$S(\theta_n, \theta_s)^2 (1 - \Theta_i) = S(\theta_i, \theta_s)^2$$

Also, as $C_n \rightarrow 1$, $(K_s - K_i)/(K_s - K_n) \rightarrow 1$. This means that for highly nonlinear soils Equation (6.10) is identical to the Equation (6.8) (with $K_i = 0$), which was originally obtained from an approximate integral form of the general Richards equation, as shown above. In the special case that $\Theta_i = 0$, or equivalently that initial water content θ_i is the same as the value θ_n where $K(\theta)$ is minimum, this equation (6.7) was shown by Broadbridge and White [1987] to follow directly from the analytically solvable flow model in the limit as $C_n \rightarrow 1$. The current generalisation to $\Theta_i > 0$ shows that for highly nonlinear soils, the original assumption $\theta_i = \theta_n$ by [Broadbridge and White, 1988] leads to no error in the prediction of time to ponding.

At the other extreme, as $C_n \rightarrow \infty$, using the scaled ponding time t_{p*} from Chapter 4 based on t_{grav}, we have

$$r_*^{-1/2}\left[1-\Theta_i exp\left\{\frac{4}{\pi}\left(\Theta_i^2-r_*\right)_{p*}\right\}erfc\left\{\frac{2}{\sqrt{\pi}}\Theta_i t_{p*}^{1/2}\right\}\right]= \\ = erf\left\{\frac{2}{\sqrt{\pi}}r_*^{1/2}t_{p*}^{1/2}\right\} \tag{6.11}$$

In the case $\Theta_i = 0$, this agrees with the explicit prediction associated with the Burgers model and found by [Clothier *et al.*, 1981]:

$$t_{p*} = \frac{\pi}{4} r_*^{-1} inverf^2\left(r_*^{-1/2}\right) \tag{6.12}$$

However unlike in the limit $C_n \rightarrow 1$, the dependence of expression (6.11) on initial water content is not due merely to the rescaling of sorptivity, which in the case of constant diffusivity, gives $S(\theta_n,\theta_s)(1 - \Theta_i) = S(\theta_i,\theta_s)$ and

$$t_p (K_s - K_n)^2 / S(\theta_i,\theta_s)^2 = r_*^{-1}\, inverf^2 (r_*^{-1/2}) \tag{6.13}$$

Equation (6.13) is the approximation that is obtained from the Burgers flow model by fitting a slightly different quadratic function $K(\theta)$ for each initial water

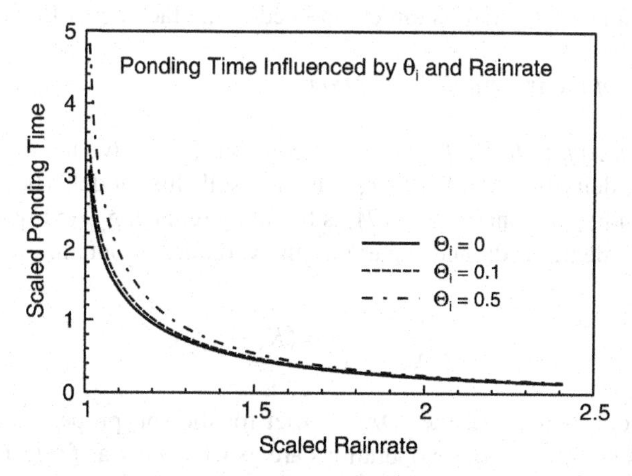

Figure 6.2. Ponding times as functions of rainrate, for the Broadbridge-White soil, by the approximate Equation (6.11), for higher Θ_i, and by Equation (6.13) for $\Theta_i = 0$.

content so that K is minimum at $\theta = \theta_i$. The differences between the predictions of ponding times given by Equation (6.13), as represented by the case of zero initial normalized water content, and by the more accurate Equation (6.11), which applies for other values of θ_i, are shown in Figure 6.2.

For such weakly nonlinear soils that can be represented by very large values of C_n, there is a noticeable effect of initial water content on dimensionless time to ponding. If the initial water content is such that the initial hydraulic conductivity is negligible compared to the saturated conductivity, then Eq. (6.13) need not be modified. However, if the initial hydraulic conductivity is non-negligible, then this will result in an increased dimensionless ponding time. This may partly explain the observed discrepancy between field data and theoretical ponding times that are typically under-predicted at high irrigation rates. However, for highly nonlinear soils represented by values of C_n close to one, as is expected to be the case for recompacted laboratory soils, the scaling within Equation (6.10) already fully accounts for the effect of initial water content.

Since Equation (6.13) has been shown to agree well with measurements on field soils, it is helpful to point out that the right hand side of Equation (6.13), that applies to weakly nonlinear soils, is closely approximated by 1.3 times the simpler right hand side of Equation (6.10) [or (6.8)], that applies to extremely nonlinear soils [Broadbridge and White, 1987]. In fact, as r_* approaches infinity and 1 respectively, the ratio of these two predictions of ponding time approaches $\pi^2/8$ (=1.23...) and $\pi/2$ (=1.57...), with the deviation from the former value pronounced only for $1 < r_* < 2$. It is remarkable that a soil's ponding time seems to be almost determined by its sorptivity and conductivity range, but is influenced little by its degree of nonlinearity. This is why, in the absence of detailed hydraulic data, it is not unreasonable to propose a universal approximation such as Equation (6.10) or (6.8) with the right hand side multiplied by 1.1 [Broadbridge and White, 1987]. This small correction is directly related to the factor b of Equation (5.21).

Other Forms of the Infiltrability Relations

The Green-Ampt Model. Equations (5.32) and (5.34) give the $\boldsymbol{I}'(f_c)$ and $f_c(\boldsymbol{I}')$ forms of this delta-function D or step-function soil flux model. The relationship between $\boldsymbol{I}'$ and t [for condition (3.17)] is found by replacing f_c in Equation (5.32) with its equivalent, $d\boldsymbol{I}'/dt$, and separating the variables to obtain:

$$\frac{I'\,dI'}{G\Delta\theta_{si} + I'} = (K_s - K_i)dt \tag{6.14}$$

We know from theory that the IDA is exact for the soil properties assumed in deriving this relation, so these operations are exact as long as $f = f_c$. For the relation with fixed $\theta = \theta_s$ [or $\psi = 0$.] upper bound, the integration limits are 0 and t, with t = 0 at $I = 0$, which produces:

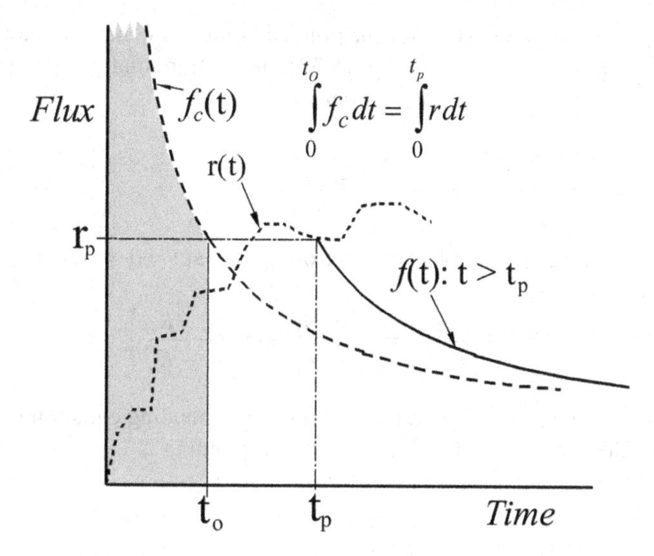

Figure 6.3. Definition diagram for the infiltrability function applied to an arbitrary rainfall pattern in the time domain.

$$(K_s - K_i)t = I' - G\Delta\theta_{si} \ln\left(1 + \frac{I'}{G\Delta\theta_{si}}\right) \tag{6.15}$$

Equation (6.4) with (6.15) will describe ponding at a time t = t_p with $\boldsymbol{I} = \boldsymbol{I}_p$. To describe the infiltrated depth after at ponding, the integration of Equation (6.14) begins at ponding, and the result is:

$$(K_s - K_i)(t - t_p) = I - I_p - G\Delta\theta_{si} \ln\left(\frac{G\Delta\theta_{si} + I'}{G\Delta\theta_{si} + I'_p}\right) \tag{6.16}$$

Equation (6.16) may also be obtained by subtraction using t_p and $\boldsymbol{I}_p$ in Equation (6.15). However, it is important to understand, referring to Figure 6.3, that when working in the time domain in the case of rainfall boundary conditions, $\boldsymbol{I'}$ is accumulated by rainfall prior to ponding and will not obey Equation (6.15) or any other $\boldsymbol{I}(f_c)$ model until after ponding. Thus Equation (6.15) must be used at or after $\boldsymbol{I'}_p$. Ponding, as stated earlier, will occur when the area under the rainfall rate curve matches the depth $\boldsymbol{I'}_p$ for infiltrability at the current rainrate, shown shaded in Figure 6.3. Note that in this figure, the symbol f_c is used to denote the infiltrability prior to ponding, but in this period it represents only a potential value based on unlimited flux at that time, and not the actual infiltration rate.

Likewise, a relation for $t(f_c)$ for the ponded boundary condition may be found by setting $I' = \int(f_c - K_i)dt$ in Equation (5.32), and differentiating both sides with respect to time, to obtain the differential equation:

$$f_c - K_i = \frac{G\Delta\theta_{si}(K_i - K_s)}{(f_c - K_s)^2}\frac{df_c}{dt} \tag{6.17}$$

This may be rearranged and integrated from $f_c = \infty$ at $t = 0$ to obtain:

$$\frac{t(K_s - K_i)}{G\Delta\theta_{si}} = \frac{K_s - K_i}{f_c - K_s} - \ln\left(\frac{f_c - K_i}{f_c - K_s}\right) \tag{6.18}$$

which applies under ponded conditions. The corresponding equation that applies in the time domain for rainfall conditions after ponding is

$$\frac{t - t_p}{G\Delta\theta_{si}} = \frac{1}{f - K_s} - \frac{1}{r_p - K_s} - \frac{1}{(K_s - K_i)}\ln\left[\frac{(f - K_i)(r_p - K_s)}{(r_p - K_i)(f - K_s)}\right] \tag{6.19}$$

The Smith-Parlange Model. Expressions for $t(f_c)$ and $t(I')$ are obtained in exactly the same manner from Equation (5.37) or (5.41). Replacing f_c with $dI'/dt + K_i$ and rearranging Equation (5.41) yields

$$\left[1 - e^{-\frac{I'}{G\Delta\theta_{si}}}\right]dI' = (K_s - K_i)dt \tag{6.20}$$

which is integrated from $t = 0$ and $I' = 0$ to obtain

$$(K_s - K_i)t = I' - G\Delta\theta_{si}\left[1 - e^{-\frac{I'}{G\Delta\theta_{si}}}\right] \tag{6.21}$$

Likewise, a relation of f_c to time may be found from Equation (5.37) by replacing I' with $\int(f_c - K_i)\,dt$, and differentiating with respect to time, to obtain:

$$f_c - K_i = \frac{G\Delta\theta_{si}(K_i - K_s)}{(f_c - K_i)(f_c - K_s)}\frac{df_c}{dt} \tag{6.22}$$

Note the similarity, except for one term, with Equation (6.17). This expression is integrated then from $f_c = \infty$ and $t = 0$ to obtain the time and f_c relation:

$$t(K_s - K_i) = G\Delta\theta_{si}\left[\ln\left(\frac{f_c - K_i}{f_c - K_s}\right) - \frac{K_s - K_i}{f_c - K_i}\right] \tag{6.23}$$

From this, as for Equation (6.19) above, the infiltrability under a flux boundary condition, after ponding at a time t_p when flux is r_p, is obtained by integrating from this point or by subtraction. In either case Equation (5.38) is obtained, which is repeated here in rearranged form:

$$(t - t_p) = G\Delta\theta_{si}\left[\frac{1}{K_s - K_i}\ln\left(\frac{(r_p - K_s)(f_c - K_i)}{(f_c - K_s)(r_p - K_i)}\right) - \frac{1}{f} + \frac{1}{r_p}\right] \tag{6.24}$$

The Three Parameter Equation As for the other relations, the three-parameter expression for $I'(f_c)$, Equation (5.47) may be directly inverted to obtain a relation for $f_c(I')$:

$$f_c = K_s + \frac{\gamma(K_s - K_i)}{\exp\left(\frac{\gamma I'}{G\Delta\theta_{si}}\right) - 1} \tag{6.25}$$

Using a method similar to those described above, the time - depth - flux interrelations for this model may be derived from Equation (5.47). However, in Equation (5.47) when I' is replaced by $\int f_c\,dt - K_i t$, and differentiated, a form similar to Equations (6.17) and (6.21) is obtained as expected, but which is difficult to integrate. In this case, it is simpler to use Equation (6.25), replacing f_c with $(dI'/dt + K_i)$, and integrating, which after some algebra reduces to an expression for I' and t:

$$I' - G\Delta\theta_{si}\ln\left[1 + \frac{1}{\gamma}\left(\exp\left\langle\frac{\gamma I'}{G\Delta\theta_{si}}\right\rangle - 1\right)\right] = (K_s - K_i)t(1-\gamma) \tag{6.26}$$

When the $I'(f)$ relationship of Equation (6.25) is substituted into Equation (6.26), the t(f_c) relation is obtained:

$$(K_s - K_i)t(1-\gamma) = G\Delta\theta_{si}\left\{\frac{1}{\gamma}\ln\left(\frac{f_c - K_s + \gamma(K_s - K_i)}{f_c - K_s}\right) - \ln\left(\frac{f_c - K_i}{f_c - K_s}\right)\right\} \tag{6.27}$$

Note that for all the forms Equations (6.18), 6.23), and (6.27), the relationship of time to f_c is implicit, and not invertable to obtain a time-explicit form. Time explicit forms will be treated below.

Table 6.1 Summary of Analytic Infiltration Functions in Normalized Form. Dimensioned equation numbers are shown in brackets.

Model	Functional Relation			
	$f_*(I_*)$[1]	$t_*(f_*)$	$I_*(f_*)$	$I_*(t_*)$
Green-Ampt ($\gamma \to 0$)	$f_* = \dfrac{1+I_*}{I_*}$ (5.34)	$t_* = \dfrac{1}{f_*-1} - \ln\left(\dfrac{f_*}{f_*-1}\right)$ (6.18)	$I_* = \dfrac{1}{f_*-1}$ (5.32)	$t_* = I_* - \ln(I_*+1)$ (6.15)
Smith-Parlange ($\gamma = 1$)	$f_* = [1-\exp(-I_*)]^{-1}$ (5.41)	$t_* = \ln\left(\dfrac{f_*}{f_*-1}\right) - \dfrac{1}{f_*}$ (6.23)	$I_* = \ln\left(\dfrac{f_*}{f_*-1}\right)$ (5.37)	$t_* = I_* - 1 + \exp(-I_*)$ (6.21)
3-parameter Combined	$f_* = 1 + \dfrac{\gamma}{\exp(\gamma I_*)-1}$ (6.25)	$t_*(1-\gamma) = \dfrac{1}{\gamma}\ln\left(\dfrac{f_*-1+\gamma}{f_*-1}\right) - \ln\left(\dfrac{f_*}{f_*-1}\right)$ (6.27)	$I_* = \dfrac{1}{\gamma}\ln\left(\dfrac{f_*-1+\gamma}{f_*-1}\right)$ (5.47)	$t_*(1-\gamma) = I_* - \ln\left[\dfrac{\exp(\gamma I_*)-1+\gamma}{\gamma}\right]$ (6.26)
Modified Philip	$f_* = \dfrac{\sqrt{2I_*+1}}{\sqrt{2I_*+1}-1}$ (5.29)	$f_* = 1 + \sqrt{\dfrac{1}{2t_*}}$ (5.28)	$I_* = \dfrac{1}{2}\left[\left(\dfrac{f_*}{f_*-1}\right)^2 - 1\right]$	$I_* = t_* + \sqrt{2t_*}$ (5.27)
Absorption[2] (exact)	$f_* = \dfrac{1}{I_*}$ (5.5)	$f_* = \sqrt{\dfrac{1}{2t_*}}$ (3.14)	$I_* = \dfrac{1}{f_*}$ (5.4)	$I_* = \sqrt{2t_*}$ (3.13)

Notes: 1. t_* is $\dfrac{t(K_s - K_i)}{G(\theta_s - \theta_i)}$ or $t\dfrac{2(K_s - K_i)^2}{S^2}$; f_* is $\dfrac{f - K_i}{K_s - K_i}$; and I_* is $\dfrac{I'}{G(\theta_s - \theta_i)}$ or $I\dfrac{2(K_s - K_i)}{S^2}$

2. For the absorption case, K_i is zero and is ignored in normalizing

Normalizing the Infiltrability Functions

Table 6.1 provides a reference summary of the infiltration functions developed here and in Chapter 5, for comparison and evaluation. Ignoring the usually small K_i term does not change the normalized functions shown in the table. The dimensioned relations presented above and in Chapter 5, and the scaling values of Equations (6.3) are simplified, however, when $K_i = 0$. Note that for the absorption relations included in this table, K_i is zero in all the scaling relations of Equation (6.3); in fact there can be no initial flow for the absorption case (with uniform initial conditions), regardless of the initial water content.

In Table 6.1, the close relation between the time-dependent absorption function and the modified Philip infiltration relation is notable. This is expected insofar as the Philip function is asymptotically exact at very small times: indeed, one consequence of the theory presented in Chapter 5 is that at small times f_c is proportional to $t^{-1/2}$ for all functions shown here. Similarly, all the functions exhibit a large time asymptote of $f_c = K_s$. Figure 6.4 is a graphical explanation of the relation among these various functions in the time domain. Here, the independent variable is $t_*^{-1/2}$, so the range of t_* shown here is from about 0.03 to ∞ , and time increases from right to left. For t_* up to about 0.6, all functions are parallel to the absorption case, but not asymptotic. The separation between the lines, in terms of f_*, ranges from about 0.3 to 1. This offset is the basis for some misunder-

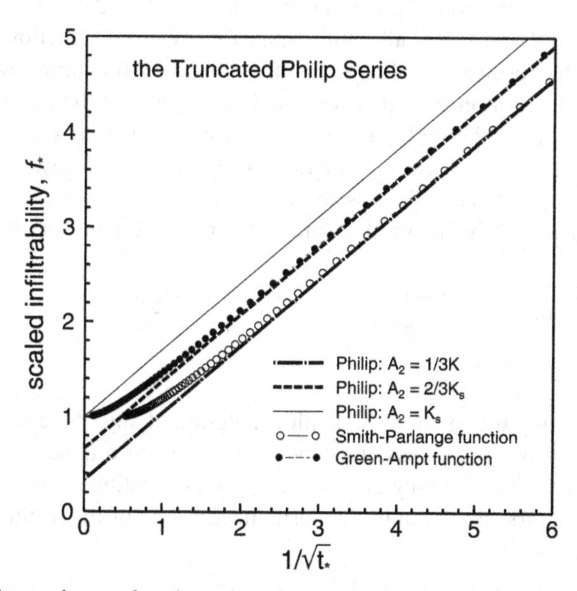

Figure 6.4. Comparisons of various time-dependent functions in the domain $t_*^{-1/2}$ illustrates clearly the differences in the short time domain. All those shown here exhibit the proper slope at short times, and approach equality at t=0, but are different by a constant factor. The Philip equation here referenced is the truncated differential of Equation (5.26)

standing regarding the truncated Philip series solution. When the term A_2 in Equation (5.26) is 2/3Ks, the early time values for this truncated series match the delta-function (Green-Ampt) expression, and $1/3K_s$ is a close match for early time values of the Smith-Parlange approximation. Use of these adjustment coefficients does not allow extrapolation to the long term, merely corrects for differences in the short time solution. Nevertheless one will often see these coefficients suggested, or used with infiltrometer data [Youngs, 1968; Hofmann et al., 2000] to estimate an unreliable value of K_s, especially if the soil under study has a short time scale, and/or there is multidimensional flow under the infiltrometer [see Chapter 8].

TIME EXPLICIT APPROXIMATIONS

Most hydrologic applications require a time distribution of infiltration rates, so the implicit functions for f_c and t and for I and t derived above may be difficult to use directly. In response, there have been attempts to find approximations to the Green-Ampt model for $t(f_c)$ that will give a function $f_c(t)$ [Li *et al.*, 1976; Stone *et al.*, 1994]. Here we will examine the derivation methods for a time-based function, and present an optimal time explicit function.

The procedure used for this purpose is the series expansion of the *ln*[-] term or the *exp*[-] term in either Equations (6.15), (6.18), (6.21), or (6.23), which relate f_c and t or $\boldsymbol{I}$ and t. Not all valid expansions result in functions which can be solved for f_c. Further, not all successful solutions will capture, with the first one or two terms of a series expansion, the key features of the relationship. The truncated form of the Philip(1957) solution , discussed above, is a case in point. The technique is more successful for Eq. (6.18) than for (6.23). The examples given below do not exhaust the possibilities.

One expansion of the natural logarithm that is convergent for all x is:

$$\ln(x) = 2\left[\frac{x-1}{x+2} + \frac{1}{3}\left(\frac{x-1}{x+1}\right)^3 + \frac{1}{5}\left(\frac{x-1}{x+1}\right)^5 + \ldots\ldots\right] \tag{6.28}$$

Clearly, use of only the first term will allow solution of any of the resulting equations when used in the expressions concerned: at worst a quadratic form for f_* will be obtained. The accuracy and success of this substitution varies, however. In the following, for clarity, only the normalized forms of the infiltrability functions will be used.

Explicit Forms from the Green-Ampt Model The substitution of the first term of Equation (6.28) may be successfully applied to Equation (6.18). If we let $u = f_* - 1$, then Equation (6.18) becomes

$$t_* = \frac{1}{u} - \ln\left(1 + \frac{1}{u}\right) \tag{6.29}$$

and use the first term of Equation (6.28) to represent the *ln*[-] term, we obtain

$$t_* = \frac{1}{u} - \frac{2}{2u+1} = \frac{1}{2u^2 + u} \tag{6.30}$$

which can be solved to obtain

$$f_* = \frac{3}{4} + \sqrt{\frac{1}{16} + \frac{1}{2t_*}} \tag{6.31}$$

In Figure 6.5 Equation (6.31) is compared to the implicit analytic model, Equation (6.18). Somewhat remarkably, given the use of only one term of the expansion, this expression captures both the short time and long time behavior that is required: df_*/dt_* at small time is -1/2, and f_* is asymptotic to 1 at large times. Moreover, this function improves on the Philip truncated series form (Equation 5.28) at intermediate times, as shown in Figure 6.5. This approximation was employed by Li *et al.* [1976].

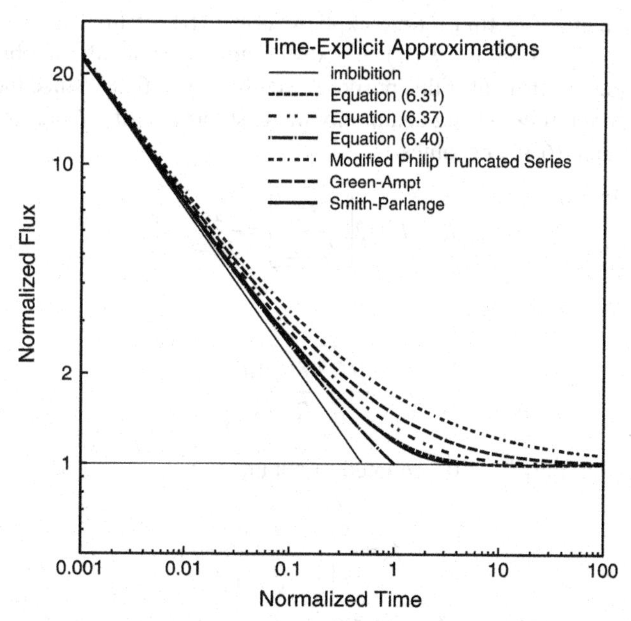

Figure 6.5. Several time-explicit approximations are obtainable by truncating a series approximation for either the log(-) or exp(-) function in the analytic infiltrability relations. Three of these are here compared with the Green-Ampt and Smith-Parlange infiltration functions.

Another series expansion of the natural logarithm is:

$$\ln(1+x) = x - \frac{1}{2}x^2 + \frac{1}{3}x^3 - .... \quad -1 < x < 1 \tag{6.32}$$

This expansion is only convergent for the limited range of x shown. If it is applied to Equation (6.29), for example, it should only be valid only for $1/u < 1$, or $f_* > 2$. Using the first two terms of expansion (6.32) in Equation (6.29), obtains

$$t_* = \frac{1}{u} - \left[\frac{1}{u} - \frac{1}{2u^2}\right] \tag{6.33}$$

which, replacing the dummy variable u, results in

$$f_* = 1 + \sqrt{\frac{1}{2t_*}} \tag{6.34}$$

which is identically the Philip truncated series infiltrability solution. It is not a good approximation to the Green-Ampt model at intermediate times, but still captures the correct long time and short time functional behavior, despite the convergence limits of the series approximation from which it is taken.

Another method to find a time-explicit form $f_*(t_*)$ relation, is to use a truncated series expansion for $\boldsymbol{I}_*(t_*)$ and differentiate the result to obtain $f_*(t_*)$. Treating Equation (6.15) for this purpose, we use series (6.28), since the range of $\boldsymbol{I}$ must go from 0 to ∞. Using again only the first term, letting $\boldsymbol{I}_* = x$ in Equation (6.28), Equation (6.15) becomes

$$t_* = I_* - 2\left[\frac{I_*}{I_* + 2}\right] = \frac{I_*^2}{I_* + 2} \tag{6.35}$$

which may be solved for $\boldsymbol{I}_*$:

$$I_* = \frac{\sqrt{t_*^2 + 8t_*} + t_*}{2} \tag{6.36}$$

This expression may be differentiated to obtain

$$\frac{dI_*}{dt_*} = f_* = \frac{1}{2}\left[1 + \frac{t_* + 4}{\sqrt{t_*^2 + 8t_*}}\right] \tag{6.37}$$

This function is shown in Figure 6.5. It is a time-explicit expression that is correct in both short and long times. If Equation (6.37) is rearranged into the following form, comparison with the other explicit approximations is clarified:

$$f_* = \sqrt{\frac{1}{16} + \frac{1}{2t_*}} + \frac{1}{2} + \frac{1}{4}\sqrt{\frac{t_*}{t_* + 8}} \tag{6.37a}$$

This demonstrates that Equation (6.37) is a slight perturbation of Equation (6.31), which is approached asymptotically at large times.

Explicit forms for the Smith-Parlange Function. With the same substitution $u = f_* - 1$, the normalized form of the Smith-Parlange $t(f_c)$ model Equation (6.20) becomes simply

$$t_* = \ln\left[\frac{u+1}{u}\right] - \frac{1}{u+1} \tag{6.38}$$

When the first term of expansion (6.28) is substituted for the *ln*[-] term in Equation (6.38), the result is

$$t_* = \frac{2}{2u+1} - \frac{1}{u+1} = \frac{1}{2u^2 + 3u + 1} \tag{6.39}$$

which obtains the normalized explicit model:

$$f_* = \frac{1}{4} + \sqrt{\frac{1}{16} + \frac{1}{2t_*}} \tag{6.40}$$

which, as seen in Figure 6.5, matches the Smith-Parlange function better than it does the Green-Ampt, but does not capture the long-time behavior of the function.

If the first two terms of the series (6.32) are used with the Smith-Parlange model, a cubic function is formed, and is not solvable for f_*. However, if only the first term of this series is used, *i.e.* representing $ln(1+x)$ by x, one obtains

$$t_* = \frac{1}{u} - \frac{1}{u+1} = \frac{1}{u(u+1)} \tag{6.41}$$

which is solved to obtain

$$f_* = \frac{1}{2} + \sqrt{\frac{1}{4} + \frac{1}{t_*}} \tag{6.42}$$

This captures the long time behavior but not the short time behavior of the function, since for these values there should be a coefficient of 2 on t_*.

Another $ln[x]$ expansion good for all values of $x > ½$ is:

$$\ln(x) = \frac{x-1}{x} + \frac{1}{2}\left(\frac{x-1}{x}\right)^2 + \frac{1}{3}\left(\frac{x-1}{x}\right)^3 + \ldots\ldots \quad x > 1/2 \tag{6.43}$$

The first two terms of this expansion may be used to produce an approximation, but it only recaptures the short time behavior (the absorption relation), Equation (3.12). Likewise, using the *exp*[-x] expansion with Equation (6.21) produces only the same short-time solution absorption function.

Generalized Explicit-time Approximations. Inspection of the results in Equation (6.31) and (6.40) suggests a general approximation, which has the correct long and short time behavior, and may be adjusted to fall between the ranges of the G-A and S-P models, as follows:

$$f_* = (1-\beta) + \sqrt{\beta^2 + \frac{1}{2t_*}} \tag{6.44}$$

For $\beta = 0$, this becomes the Philip modified function, and for $\beta = 1.0$, it falls intermediate between the Smith-Parlange and Green-Ampt functions. This is illustrated in Figure 6.6. As indicated by this generalized function, the long time behavior error of the approximate Equation (6.40) is easily corrected by changing the factor 1/16 to 9/16, or β^2. Similarly, the short time behavior of Equation (6.42) is easily corrected by changing the term t_* to $2t_*$. Indeed the success in fitting with such approx-

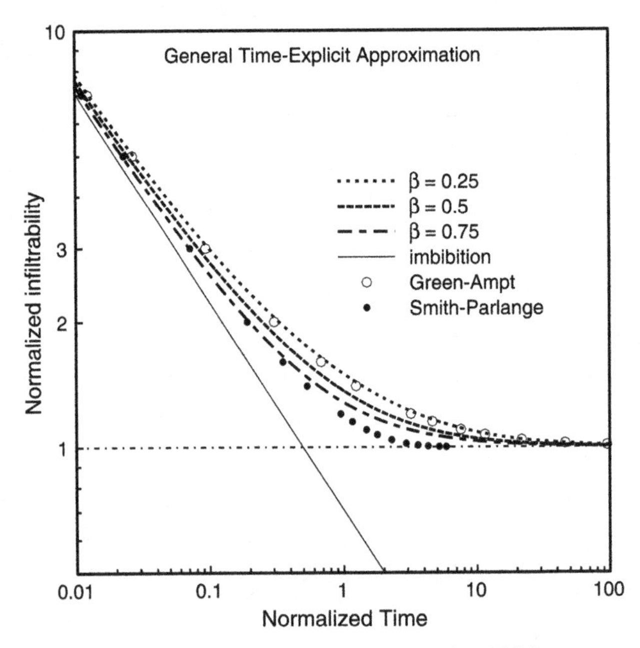

Figure 6.6. The approximate time-explicit expression Equation (6.44) can represent the general 3-parameter infiltrability function reasonably well over the entire time scale by judicious choice of parameter β.

imations and the improvement due to the correction term in Equation (6.37a) suggests there may be innumerable ways to fit time-explicit functions to these normalized models, if desired, without the addition of extra empirical parameters. Such a correction term approach for the $I_*(t_*)$ relation was used by Stone *et al.* [1994] in improving a truncated series approximation to the Green- Ampt model.

An alternate and apparently even better generalized time-explicit approximation can be obtained from Equation (6.37a). Following the method for the generation of Equation (6.44) the coefficients $(1/16)^{1/2}$ and 1/2 are made parameters. However, the sum of the terms must satisfy the long-time asymptote of $f_c{}^* = 1$. With this constraint, we obtain:

$$f_* = \sqrt{\omega^2 + \frac{1}{2t_*}} + \left(\frac{3}{4} - \omega\right) + \frac{1}{4}\sqrt{\frac{t_*}{t_* + 8}} \tag{6.45}$$

Values of ω between 0 and 0.75 are possible without violating either asymptotic condition. Figure 6.7 illustrates this function with ω taking values of 1/8, 1/4, and ½. The use of ω = ½ is an excellent approximation for the Smith-Parlange model, and ω = 1/8 is an equally close approximation for the Green-Ampt function. In neither case have these values of ω been optimized.

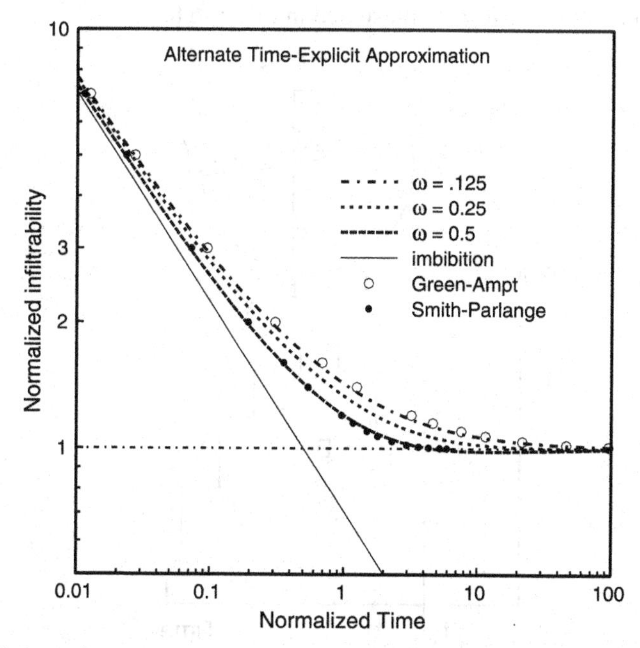

Figure 6.7. A more general time-explicit form, Equation (6.45), with ω between approximately 1/8 and ½ provides an even more accurate approximation to the infiltrability function than does Equation (6.44), as illustrated here.

APPLYING THE INFILTRABILITY MODELS TO VARIABLE RAINFALL PATTERNS

The value of the IDA type models developed above is, among other things, the robust applicability to predict runoff from rainfall patterns typically found in natural rainfalls. Probably the chief practical obstacle to the use of infiltration theory for runoff problems, in fact, is the unfortunate paucity of records of the distribution of rainfall intensities for storms of hydrologic importance. This lack of important rain rate data is usually a result of economics, since rainfall rate data, which infiltration theory demonstrates is important, is significantly more expensive to collect than simple daily rainfall totals.

When rainfall rate records are available, they generally consist as a time series of fixed rates lasting a short interval of time. This derives from the nature of the data collection methods, whether they be weighing rain gages or tipping-bucket digital recorders. Continuous weighing methods create an analog graph of cumulative depth vs. time, which is divided into a series of slope segments in the interpretation into rates. The tipping bucket recorders create either a series of times at which a certain amount has been accumulated and has caused the bucket to tip, or else a record of uniform time intervals in which the total number of tips per interval is recorded (including zero). In all cases, a record will appear as stepwise constant rates, such as illustrated in Figure 6.8.

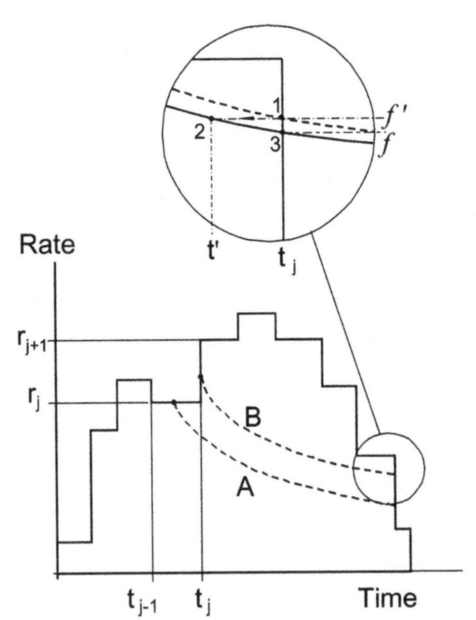

Figure 6.8. Definition diagram for the calculation of infiltrability during a rainfall record as recorded using a tipping bucket or weighing raingage. Curves A and B are two pond occurrence modes, described in the text.

Equations (6.4), (6.5), or (6.6) assume a uniform rainfall rate. With the IDA, use of cumulative depth rather than time will allow application of infiltration theory to variable rain rates. The soil will at first accumulate an infiltrated depth $\boldsymbol{I}$ following the rainfall record, which we will call $\boldsymbol{I}_a(r)$. The infiltrability relation $\boldsymbol{I}(f_c)$ will describe a ponding depth, as discussed above, when the current rainfall rate, r, is substituted for f_c as in Equations (5.32), (5.37), or (5.47). This relationship calculates a value of $\boldsymbol{I}_p(r)$, were the current rate r to continue to ponding. In the early stages of any rainfall, necessarily $\boldsymbol{I}_a < \boldsymbol{I}_p$. Here we assume that the rainfall rate is generally greater than K_s, the minimum for ponding to occur, and for simplicity we treat the case where K_i is negligible. In the next chapter, the small r case and the significant K_i cases will be treated.

Finding the Ponding Time The value of $\boldsymbol{I}_a$ will increase steadily, while the value of $\boldsymbol{I}_p$ is always a function of and inversely related to the current rate r(t). Given the stepwise type of record, the ponding time or value when the $\boldsymbol{I}_p(r)$ equation is just satisfied by $\boldsymbol{I}_a$ can occur in two ways:

a) The ponding time may occur during a time step at a given value of r as $\boldsymbol{I}_a$ is increasing at the rate r. This is shown as infiltrability curve A in Figure 6.8.

b) Alternatively, since these equations have $\boldsymbol{I}_p$ decreasing with larger r, the infiltrability equation may be satisfied at the point at which r increases to the value at the next interval: curve B in Figure 6.8.

Case (a): The calculation method in either case must keep track of $\boldsymbol{I}_a$. At the end of any step, j, we have:

$$I_a(j) = \sum_{k=0}^{j} r_k dt_k$$

During any step j of rate r_j, the cumulated rainfall depth at the end of the step, $\boldsymbol{I}_a$(j) is compared with the infiltrability value from the $\boldsymbol{I}_p(f=r)$ equation. When $\boldsymbol{I}_a(j-1) < \boldsymbol{I}_p(r_j) < \boldsymbol{I}_a(j)$, ponding must occur within the step. Ponding time is simply calculated as follows:

$$t_p = t_{j-1} + \frac{I_p(r_j) - I_a(j-1)}{r_j} \tag{6.46}$$

Case (b). Whenever the rainrate increases from r(j-1) to r(j) between intervals, $\boldsymbol{I}_p(r_j)$ is compared with $\boldsymbol{I}_a$(j-1) prior to calculating $\boldsymbol{I}_a$(j). When $\boldsymbol{I}_p(r_j) < \boldsymbol{I}_a(j-1) < \boldsymbol{I}_p(r_{j-1})$, then the ponding depth is $\boldsymbol{I}_a$(j-1) and ponding occurs during the increase in rainrate at the end of interval (j-1). The infiltrability at ponding, $r_{j-1} < f_p < r_j$, is calculated using the selected infiltrability relation $f_c(I)$ - *e.g.* Equations (5.34), (5.41), or (6.22):

$$f_p = f_c(I_a[j-1])$$

These calculations can be done in computer code or in most cases with a spreadsheet program.

Tracing the Infiltrability Curve. Following the determination of ponding time, t_p, the infiltrability curve may be followed through the time period of the rainfall record until rainfall rate falls below the infiltrability. In hydrologic computations, the objective is to check the infiltrability against rainfall rate during each time step through the rainfall record, and to calculate the rainfall excess rate (the rainfall not infiltrating and potentially creating runoff) during that interval. The implicit nature of the $t(f_c)$ relations in Table 6.1 has distinct disadvantages in this calculation compared with time-explicit formulations, and has often caused practicing hydrologists to choose more empirical formulae. This is the background for the search for time-explicit forms described above. If infiltration is to be part of a computer hydrologic model, simple implicit solution methods exist to deal with the infiltrability equations. The direct solution option in such a case requires iterative solutions. Newton-Raphson iteration can be used to find the f_c for any and all rainfall rate change time steps following t_p. For this purpose the forms of the solution such as Equation (6.19) and (6.24) involving both t and t_p may be used directly as soon as t_p and r_p are found.

As a second option, the time-explicit approximations given above [e.g. Equation 6.45] may be adopted and used to estimate the value of f_c for any time until the runoff ceases, *i.e.* while $f \leq r$. This choice will depend on the desired accuracy: for most practical applications this should be sufficient. For the explicit method, the time variable used in the approximate equations, such as Equations (6.44) or (6.45), must include an adjustment such that $f = r_p$ at the time $t = t_p$. One cannot use $t - t_p$ in the time explicit forms, as in Equation (6.24), because that is an implicit difference equation determined by subtraction. Rather for following the infiltrability curve $f_c(t)$ one must first solve for a time t_a which is the time that an infiltrability curve would have started in order to pass through the point (r_p, t_p). Then the adjusted time to use in subsequent infiltrability calculations is $t' = t - t_a$ ($t > t_p$). Referring to Figure 6.3, $t_a = t_p - t_o$. By definition, $t' = t(f_c = r_p)$ at $t = t_p$. If the calculations are to be made in normalized terms, then the adjusted time variable t' is normalized the same way. Thus all rainfall times following t_p should be adjusted in this manner for computation purposes.

A third option for more accurately calculating the time evolution of infiltrability is a simple 3 step procedure that can be used to maintain the accuracy of the infiltrability equations without complete iterative solutions. For this example, assume one has chosen to use the Green-Ampt formulas: For each interval, j, after t_p, the calculation steps are as follows:

1. At the adjusted time, t'(j), for which the infiltrability $f_c(j)$ is needed, use an explicit relation [such as Equation (6.45) with $\omega = 0.5$] to make a first estimate of the value of $f_c(j)$, which may be called f_1.

2. Use f_1 in an implicit relation [such as Equation (6.18)] to find the correct time t_1' for this infiltrability, finding a small time correction $\Delta_e t = t'(j) - t_1'$.
3. Use the slope of the infiltrability curve to make the small correction in f_1 that is indicated. The slope $f'(t')$ may be very accurately estimated by differentiating the time explicit Equation (6.45). The resulting expression is

$$\frac{df_*}{dt_*}\left(t_*\right) = \frac{\left(t'_* + 8\right)^{-3/2}}{\sqrt{t_*}} - \frac{t'^{-3/2}_*}{\sqrt{16\omega^2 t'_* + 8}} \tag{6.47}$$

This expression is graphically compared with the value of df_*/dt from Equation (6.17) in Figure 6.9. An expression for df_*/dt can alternatively be obtained by differentiating the $\boldsymbol{I}(f_c)$ expressions, *e.g.* Equations (5.32) or (5.37), which obtains df/dt as a function of f_c. These are given above as Equations (6.17) and (6.22). In either case a very accurate value for $f(j)$ at $t'(j)$ may now be calculated with the small correction as follows, using $\Delta_e t$ from Step 2 above:

$$f(j) = f_1 + \left[\Delta_e t\right]\frac{df}{dt'} \tag{6.48}$$

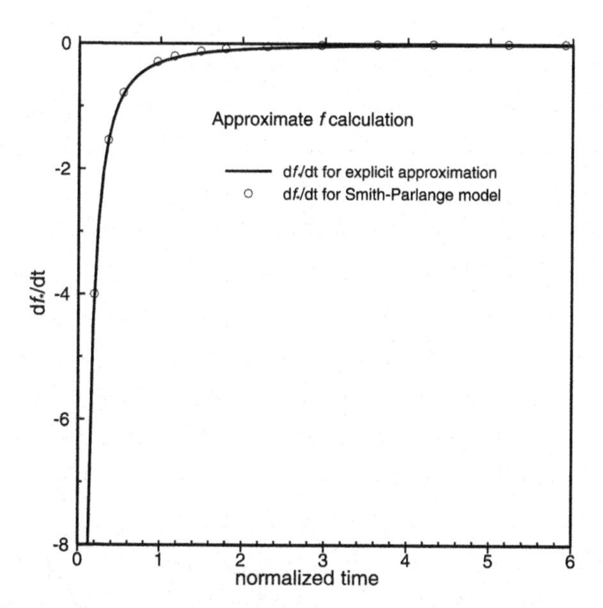

Figure 6.9. The time explicit approximation Equation (6.47) is practically as good as the more exact Equation (6.17) for use in the prediction-correction method for infiltrability calculations with variable rainfall rates, Equation (6.48).

One may choose to work in dimensional units or with the normalized values of f_c and (adjusted) time. The computational steps are illustrated in the inset to Figure 6.8.

SUMMARY

Three infiltration models arise from the approximations introduced in Chapter 5, and all forms of the relations between f_c, $\boldsymbol{I}$, and t are presented here in normalized form. Time of ponding expressions are one means of comparing the performance of all the models derived above. Several time explicit approximations have been introduced based on truncated series approximations for *ln*(•) or *exp*(•) functions contained in the three models. Two of these are notable in their accuracy of approximation, and useful in hydrologic models, especially irrigation infiltration.

Finally, several alternate computational procedures have been suggested for calculation of infiltrability during a rain of stepwise varying rates. In the next chapter the application of these models in cases of natural complexity of soils and rain patterns is introduced.

7

Applying Infiltration Models in Layered Soils and Redistribution Cases

INTRODUCTION

In the previous chapters, infiltration models have assumed a uniform soil profile. It is understood that this in fact is not always the case: especially for cultivated soil; the rainfall often finds a loose, tilled soil and subsequently rainfall energy creates a surface layer of greater density and lower K_s than that deeper in the soil. In addition, there may be soil layers near the surface that affect infiltration and should be dealt with in infiltrability calculations.

The application of infiltration models employing the IDA was demonstrated with patterns of rainfall composed of a sequence of stepwise uniform rates, as is typically obtained from recording equipment. In the previous chapter, these rates were assumed to be composed of a cluster in which all $r > K_s$. The patterns of rainfall often contain extended periods of rates lower than K_s, or zero rates, which may be early in the storm or may occur in the middle of a storm, creating a hiatus during which potential infiltrability recovers as the water redistributes within the upper soil, or the soil dries through evaporation.

This chapter presents methods for dealing with these complications using the infiltration theory presented earlier, with the addition of other assumptions that will be used when necessary. We first deal with a method for simple redistribution between inputs, then infiltration for a layered soil, and finally a model to treat all these conditions [Corradini et al., 2000].

REDISTRIBUTION AND REINFILTRATION

For hydrologic applications, infiltration theory needs to be extended to cases where rainfall ceases or is significantly reduced within a storm period. Thus one

Infiltration Theory for Hydrologic Applications
Water Resources Monograph 15

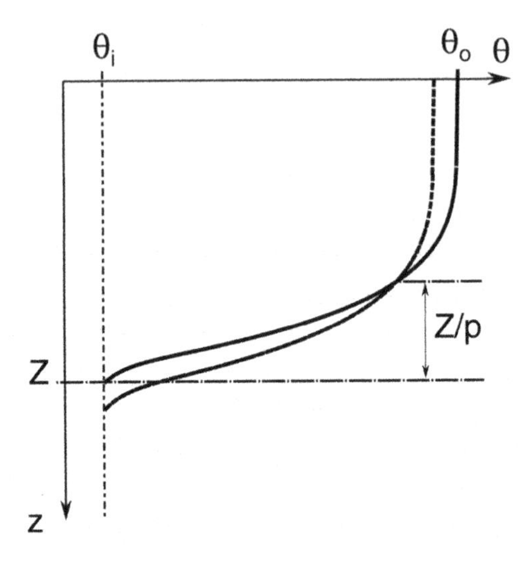

Figure 7.1. Definition sketch for the change in the wetting profile after the surface input r is reduced to less than K_s: redistribution conditions. The profile continues to extend downwards but the water content at the surface recedes, as shown by the dotted line.

must calculate both redistribution and reinfiltration conditions. The latter case requires the former, since reinfiltration occurs through a wetted zone near the surface that may be encompassed by the wetting of the infiltrating water. This requires some estimate of the depth of wetting and water content near the surface at the end of the redistribution period.

Redistribution of Water During Rainfall Hiatus

At the end an infiltration event, an idealized schematic of the water content profile in a soil might be as illustrated in Figure 7.1. An amount $\boldsymbol{I}$ that has infiltrated may be represented by a depth Z that has mean water content $(\theta_o - \theta_i)$. The profile need not be piston like, but we rather assume that the profile shapes are similar functions, $\theta(z)$ containing volume $\boldsymbol{I}$, where $\boldsymbol{I} = \beta Z(\theta_o - \theta_i)$, and β is defined as

$$\beta \equiv \frac{1}{Z(\theta_o - \theta_i)} \int_0^Z (\theta(z) - \theta_i)dz \tag{7.1}$$

β will be expected to lie between about $\pi/4$ (a quarter-elipse) and 1.0 (a rectangle). Redistribution is assumed to consist of a lengthening of the profile due to capillary and gravitational gradient at the advance edge, with small additions or losses at the surface, balanced by reductions in the 'width' of the profile, $\theta_o - \theta_i$. Designating t_e as the time when the rainfall event ended and the redistribution period begins, with small or negative surface flux v_o, the mass balance is represented mathematically as

$$\beta Z(\theta_o - \theta_i) = I' + (v_o - K_i)(t - t_e) \tag{7.2}$$

The situation is diagramed in Figure 7.1, with the profile changing from the solid to the dotted line in a given time step. This balance statement can be also represented in differential form as follows:

$$(\theta_o - \theta_i)\frac{dZ}{dt} = \frac{v_o - K_i}{\beta} - Z\frac{d\theta_o}{dt} \tag{7.3}$$

The profile extension represented by the left side of this expression can be treated like the piston flow advance in the Green-Ampt model. The difference is that the capillary drive and the conductivity are less than that for saturation, and are reduced appropriately as θ_o declines during elongation. In Chapter 2 we defined a generalized flux potential, $\phi_{a,b}$ that can be written also in terms of water content, $\phi(\theta_o,\theta_i)$:

$$\phi(\theta_i,\theta_o) \equiv \phi_{i,o} = \int_{\theta_i}^{\theta_o} D\,d\theta = \int_{\psi_i}^{\psi_o} K\,d\psi \tag{7.4}$$

Note that ϕ may be thought of as a function of ψ as well as θ. We use this definition along with a gravity term, $K(\theta_o)$, to express the approximate Darcy elongation represented by the left side of Equation (7.3). Referring to Figure 7.1, a point of rotation is assumed which is some fraction 1/p of the depth Z above the advance depth. Below this point θ is increasing and above it θ is decreasing. With these two definitions, the elongation term is expressed as follows:

$$(\theta_o - \theta_i)\frac{dZ}{dt} = \frac{1}{\beta}\left[K(\theta_o) + \frac{p\phi_{i,o}}{Z}\right] \tag{7.5}$$

The shape factor β is included here since it alters the relation between the elongation flux and the rate of elongation. Using Equation (7.5) in (7.3), and rearranging, one obtains:

$$\frac{d\theta_o}{dt} = \frac{\Delta\theta_{io}}{I'}\left[q - K_i - \left(K(\theta_o) + \frac{\beta p\phi_{i,o}\Delta\theta_{io}}{I'}\right)\right] \tag{7.6}$$

Here, Z has everywhere been replaced by $I/\Delta\theta_{io}\,\beta$. This differential equation can be solved by standard Runge-Kutta methods. It is evident that use of this model assumes some knowledge or approximation of the variation of ϕ with water content (or ψ). Approximation of this variation is not difficult, insofar as S varies very closely as the square root of the saturation deficit, $\Delta\theta_{si}$.

Equation (7.6) is in effect a conceptual relative of the Green-Ampt piston flow model, extended to include the development of the piston prior to ponding, and the reshaping of the "piston" when rain rates fall below K_s. An expression for $\theta_0(t)$ such as Equation (5.50) may be also used to calculate the initial rising surface water content. In either model v_o may be either greater or less than K_s. Equation (7.6) has the advantage of dealing directly with stepwise changes in v_o. $K(\theta_o)$ (and ϕ_{io}) increases as θ_o increases, and K can reach the saturated value K_s whenever rainfall rates exceed K_s for sufficiently long periods. For those cases where rainfall rates rise to exceed K_s after a long initial period of lower rates, an updated value of θ_i is subsequently used, based on the previously developed profile characterized by θ_o. This will be like the reinfiltration case described below, since the extent of the depth associated with this value of initial water content is limited, and must be considered in the subsequent calculations.

Reinfiltration Calculations

Simple reinfiltration occurs when a potentially runoff-producing rainfall pulse ($r > K_s$) occurs following a redistribution period. The value of initial water content seen at the beginning of reinfiltration is the result of redistribution, and is usually a relatively large value θ_{i2}, during which period the value of $K_i(\theta_{i2})$ cannot be neglected. This elevated θ_{i2} value is effective until the reinfiltration wetting pulse depth $Z_2 = I_2/[\beta(\theta_s - \theta_{i2}]$ reaches the depth of the previous wetting depth, Z_1, at which point the original initial conditions apply again, and the two "waves" merge into one. The conditions are illustrated schematically in Figure 7.2. This is a case where the formulations given in Chapter 6 explicitly including values for K_i are required. The only unique part of reinfiltration computations is the accounting for the two depths and the transition to original conditions when Z_1 is reached. Several applications of reinfiltration approximations are included in Smith *et al.*, [1993].

LAYERED SOILS

In order to deal with the complexity of a layered soil profile, we envision a conceptual model of the wetting profile in which water moves sequentially piston- like through one layer and then the next. In fact, there is a transition as the wetting "front" moves through the soil interface. Owing to the uncertainty of variation in a profile-effective value for sorptivity or G as a soil wets through

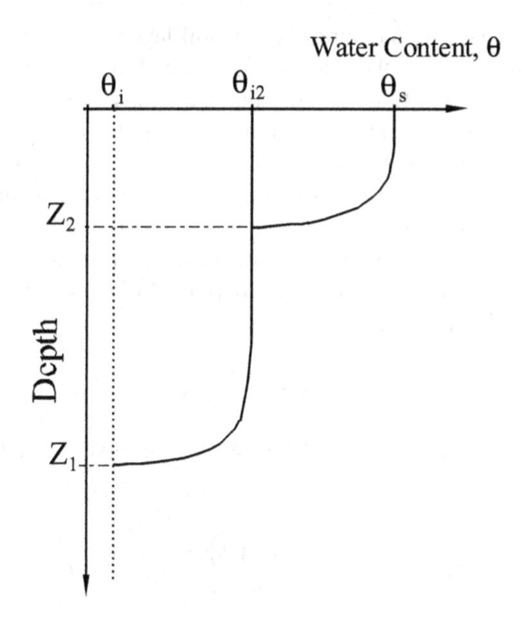

Figure 7.2. After redistribution, or after infiltration at a rate $r < K_s$, the increase in r to a value larger than K_s can create a secondary wetting pulse as shown here. The dotted line indicates the initial value of profile water content, assumed constant.

arbitrary layers, this assumption is necessary in practical terms, and should generally be adequate. Lack of knowledge of the various layer properties is far more likely to be the major source of uncertainty or error. The conceptual model for piston flow dictates that, as the wetting "piston" moves to include layers of different properties, we assume (a) that the effective composite parameter for K_s should represent the composite effective value of steady saturated flow through the profile down to the location of the piston front, and (b) the effective value of capillary drive G or sorptivity S should represent that for the wave front, where capillary gradients are concentrated. Thus these two effective values should change as the wetting front moves into a new layer. This has been demonstrated by solutions of Richards' Equation [Smith, 1990] for layered conditions.

Asymptotic f_c for General Layered Case

The theory presented above for a single soil has infiltration rates asymptotic to $K(\psi = 0) = K_s$ at large times. For layered soils, the asymptotic value for infiltration must be obtained as the steady flow rate through the profile for $\psi_o = 0$. We designate this asymptotic value as K_∞. As above, subscript o indicates surface conditions.

Assume that a profile is composed of n soil layers, k = 1,2,...n, of arbitrary thickness $\Delta z_k = z_k - z_{k-1}$, and the last layer is assumed to extend indefinitely. Each layer may have a different $K(\psi)$ relationship, which is represented by $K_k(\psi)$. For steady flow there is a fixed value of ψ at the bottom of each layer from 1 to n-1, termed ψ_k, plus a head at the surface, ψ_o. The value of ψ is continuous across each layer interface, while the values of θ are not. Figure 7.3 is a schematic illustration of the progression of successive gradients of ψ is a multiply layered profile.

At very large time the asymptotic value of infiltration rate, $v_o = K_\infty$, will be the f due to $\psi_o = 0$ (assuming no surface depth). Within each layer, $v = v_o$, and Darcy's Law will describe the steady flow:

$$v_o = -K_k(\psi)\frac{d\psi}{dz} + K_k(\psi) \tag{7.7}$$

This may be rearranged and integrated for each layer (except the last one, which extends indefinitely):

$$z_k - z_{k-1} = -\int_{\psi_{k-1}}^{\psi_k} \frac{K_k(\psi)d\psi}{v_o - K_k(\psi)} \qquad k = 1, n-1 \tag{7.8}$$

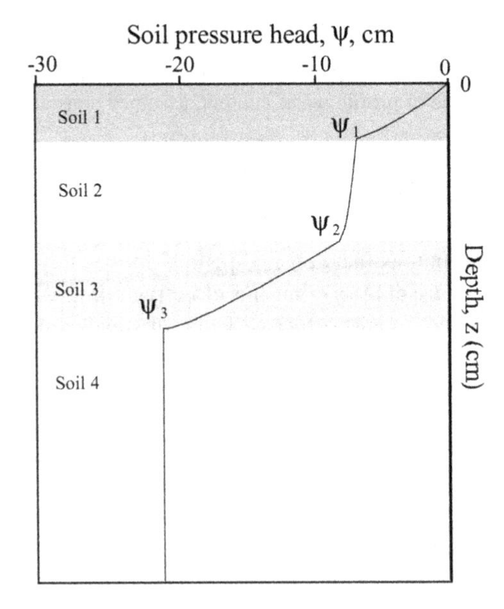

Figure 7.3. The steady flow profile in a multi-layer soil, that is the asymptotic large-time condition under infiltration, may have a variety of changes in pressure head gradients, depending on the soil hydraulic characteristics of the set of layers involved. The lowest layer here assumed to have semi-infinite depth, will have a uniform value of ψ under steady flow.

The Equations (7.8) comprise a set of n -1 equations with n unknowns. The last equation needed is the steady flow in the nth layer:

$$v_o = K_n(\psi_{n-1}) \tag{7.9}$$

Substituting Equation (7.9) into the Equation (7.8) for layer n-1 produces a set of n - 1 equations and n - 1 unknowns, which are the values of ψ_k at the bottom of layers 1 through n-1. This implicit set of equations may in principle be solved for the values of ψ_k , k = 1, n-1, for an arbitrary pattern of layer properties $K_k(\psi)$. There may not, however, always be a valid solution, depending on the array of layering. Values of ψ may become positive within intermediate layers, which is physically valid, but an equation other than Equation (7.8) must be used to reflect a constant K_s for $\psi > 0$. In any case, due to the highly nonlinear change of K with ψ, solution of the system can be difficult for multiply layered soil profiles.

As the wetting proceeds through a series of layers, the value of K_∞ to use (in place of K_s) in the infiltration equations should shift to reflect the changed layering. Thus K_∞ is a parameter value which changes as wetting progresses. The value of the effective K_∞ should be computed based on the layer number within which the profile is currently wetting. The value of G appropriate to use in a layered case must also be the one at the front of the wetting "piston", which will change as infiltration proceeds. Given the appropriate G and K_∞, the infiltration models developed in Chapter 5 and 6 may be applied as an approximation to layered soil infiltration.

For a two-layer system, such as a crust-topped soil, determining K_∞ is rather straightforward, as Equation (7.9) may be substituted into (7.8) and the resulting single equation solved iteratively for ψ_1 [Smith, 1990]. As discussed below, transitional characteristics of actual $f_c(I)$ as a wetting zone moves through a layer interface are of course not well simulated by a stepwise change in effective parameters. For larger numbers of layers, there are methods to deal with particular cases, such as steadily increasing values of K_k [Bouwer, 1969]. Moore [1981] published a method for two layers which assumes that the value of ψ_1 is (algebraically) greater than ψ_B of the Brooks-Corey hydraulic characteristics (see Figure 2.6) and thus assumes that upper layer K does not change with ψ. This allows explicit solution of Eq. (7.7) without integration, but is a dubious assumption for general use.

A Single Restrictive Upper Layer

Most approaches to infiltration through a 'crust'—a thin surface layer with lower K—have assumed the crust is sufficiently thin to be immediately wetted, and thus have constant properties. Here we treat surface layers in general, with the crust condition a special case. The models are conceptual extensions of the models presented in Chapter 6. The surface layer properties are indicated by the subscript 1, and those of the subsoil by subscript 2. Subscript b refers to the boundary or interface between the layers. Because of the transitional nature of the basic parameters, normalized versions of the infiltration models as given in Table 6.1 cannot be obtained in general without difficulty.

Smith [1990] demonstrated that the IDA holds for layered cases. The difficulty lies in reproducing the transitional shape of the infiltrability relation $f_c(\boldsymbol{I})$ as the wetting front passes through the interface. The changes to the simple $f_c(\boldsymbol{I})$ relation at that interface have direct relation to the differences in the parameters K_s and G for the two soils. Figure 7.4 illustrates the $K(\psi)$ relation for a variety of possible differences between layers. When the upper soil is a crust formed from compaction of the lower soil, one expects by physical arguments and experience, that $K_{s1} < K_{s2}$ and at the same time $\psi_{B1} > \psi_{B2}$, and $G_1 > G_2$ [Mualem and Assouline, 1989]. Mualem and Assouline [1989] argue that the crust is not a distinct layer but is composed of a transition in soil properties from the surface into the subsoil. There remains no analytic treatment for the flow dynamics in this case, however. The expected changes in the $K(\psi)$ relations, referring to Figure 7.4, would have a compacted crust from soil A with a $K(\psi)$ relationship something like curve C. In general, among various soil textures, the same is true for different compactions: larger values of saturated conductivity tend to be associated with smaller values of capillary drive G, or in the case of the TB-C type soil, a smaller parameter ψ_B [or a larger van Genuchten α_G]. This may lead to a point of intersection of the two $K(\psi)$ curves, as shown. The value of steady interface potential ψ_b cannot exceed the value that makes $K_A(\psi_b) = (K_s)_C$.

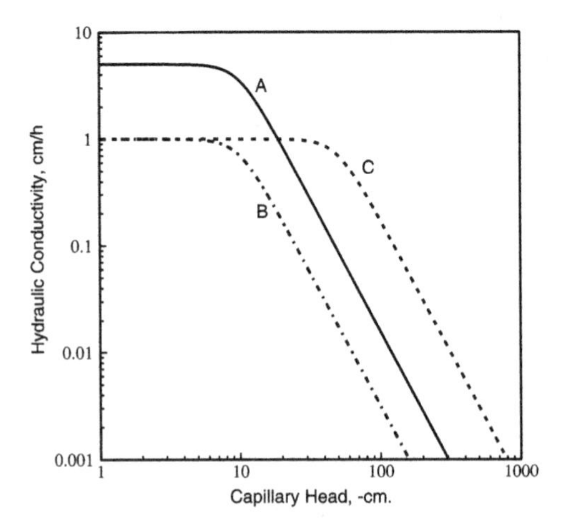

Figure 7.4. When a crust is formed at the surface of a soil profile by compaction, the soil $K(\psi)$ curve can be expected to change from one like A to one like C, with shifts in air entry parameter ψ_B and saturated conductivity.

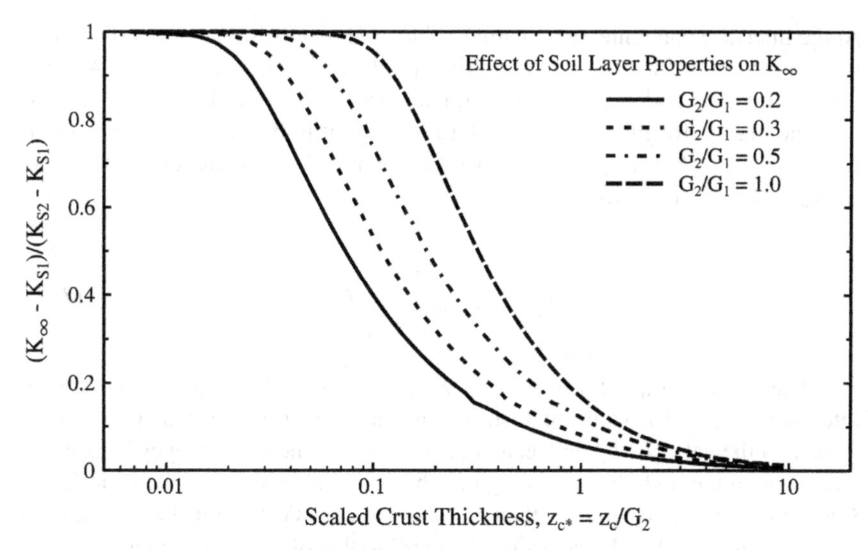

Figure 7.5. The effective value of final infiltration rate through a two-layer crust depends on the upper layer (1) depth and the ratios of Ks and G in the two layers, as illustrated here.

Let $\boldsymbol{I}_b$ represent the infiltrated depth that just fills the upper layer. For any significant thickness of the upper layer, ponding and runoff may start before water reaches the interface: $\boldsymbol{I}_p < \boldsymbol{I}_b$. The theory for uniform soil may be applied to calculate the ponding time and infiltrability as long as $\boldsymbol{I} < \boldsymbol{I}_b$. When the lower layer is encountered, the value of K_∞ that will apply to subsequent infiltration depends on the ratios of the two values of K_s but also on the ratios of G and the upper layer thickness. An overview of the effects of these various properties on K_∞ is shown in normalized terms for a variety of parameters in Figure 7.5. These curves were obtained by solution of Equation (7.8). They demonstrate that a very thin crust will have little or no effect on the value of K_∞, but this will in part depend on the shape of the $K(\psi)$ relations involved, as seen in Figure 7.4.

Layer Infiltrability Transition. Results of Smith [1990] indicate that the interface transition is more gradual for the effects of G than for change from K_{s1} to K_∞. Those extensive simulation results also indicate that infiltrability is in all cases nondecreasing with $\boldsymbol{I}$ or time. Thus when a change of soil is encountered at the interface that would otherwise increase infiltrability (which is physically unusual), the effective value of G apparently changes gradually. In part this is related to the fact that the wetting "front" is not a "piston" type wave.

One means to model this transition is to calculate what may be called infiltrability matching parameters at the time when the interface is encountered. At this point, given $K_{s1} < K_{s2}$, the new value of K_∞ always lies between the two K_s values, as shown in Figure 7.5. The infiltrability is f_{cb} at the time $\boldsymbol{I}_b$ is reached.

At the interface encounter, a matching value of G may be found, termed G_m, that in the infiltration function along with K_∞ produces a value equal to f_{cb}. Assume the infiltrability model employed is Equation (5.37), and for clarity let us take K_i to be negligible for this case. The saturation deficit in the second layer is applicable for $I > I_b$. Solving Eq. (5.37) for G, assuming $K_i = 0$, and using K, I_b and f_c, G_m would be found as:

$$G_m = \frac{I_b}{(\theta_{s2} - \theta_{i2}) \ln\left(\frac{f_c - K_\infty}{f_c}\right)} \tag{7.10}$$

A useful technique for estimating f_c as the wetting front moves beyond the interface is to use a value of G in the infiltrability function that is weighted (based on the value $I - I_b$) between G_m and the G_2 of the second layer. This being a conceptual method, the weighting may be designed in any manner such that as I becomes sufficiently greater than I_b, G approaches G_2 [Smith, 1990]. One can in a like manner, when I reaches I_b, change G to that of G_2 and calculate a matching transitional value of K_{sm} that approaches the value K_∞, again by weighting, during further increases of I beyond I_b. While rather tedious for hand calculation, this is a simple enough calculation approach for computer simulation.

Infiltration and Redistribution/Reinfiltration in Layered Soils

Methods for dealing with both layered soils and reinfiltration or redistribution were described above, which are consistent with the analytical models derived in Chapter 5 and the overall IDA model theory. A conceptual model has recently been developed [Smith *et al.*, 1999; Corradini *et al.*, 2000] that deals with all the above complications to simple infiltration theory. The model has been successfully validated against Richards' equation solutions for various combinations of soil property changes and layer thicknesses. This model will be outlined here.

For treating redistribution or infiltration in a layered system, the water content, which is a convenient variable for a uniform soil (being linked more directly to I), cannot be used for the layered case, because it is the capillary head that is continuous across a soil interface, rather than the water content. At the interface there is one water content for the upper soil, θ_{1b}, and another in the lower soil, θ_{2b}. Two variables in the two-layer system must be found simultaneously, that describe the conditions at the surface and at the interface. An additional equation describes the continuity of flow across the interface.

The uppermost of the two-layer system in the Smith/Corradini model need not be a thin crust. As illustrated in Figure 7.6, the layer is conceptually partitioned by a partition coefficient, α. For this two-layer model, the state variables are the infiltrated depth in the upper and lower soils, I_1 and I_2, and there are two unknowns, the capillary head at the surface, ψ_o, and the capillary head at the interface, ψ_b. Prior to infiltration reaching the interface, when $Z < z_b$, the model acts like a single layer model. Corradini *et al.* [2000] and Smith *et al.* [1999] used the partition coefficient α as a means to distinguish between the effect that infiltration flux at the surface, v_o, has on the surface water content, θ_o, and the interface water content, θ_{1b}, after the advancing wetting front reaches the interface. The flux through the interface, v_{12}, is found as the remainder from the flux from the surface, v_o, after an amount has been subtracted that goes to wetting the layer above the interface, thus:

$$v_{12} = v_o(\psi_b, \psi_o) - (1-\alpha)Z_b \frac{d\theta_{1b}}{dt}$$
$$v_o(\psi_b, \psi_o) = \frac{\phi_1(\psi_b, \psi_o)}{Z_b} + K_o \tag{7.11}$$

Note that the flux potential ϕ is now expressed as a function of the capillary potentials rather than water contents. Some knowledge of the variation of ϕ with ψ is required, as well as an approximation for the retention relations for the two soil layers. Given that, it is a far simpler solution than a numerical solution of Richard's equation, and probably more appropriate to the uncertainty of soil hydraulic properties typically accompanying hydrologic simulation.

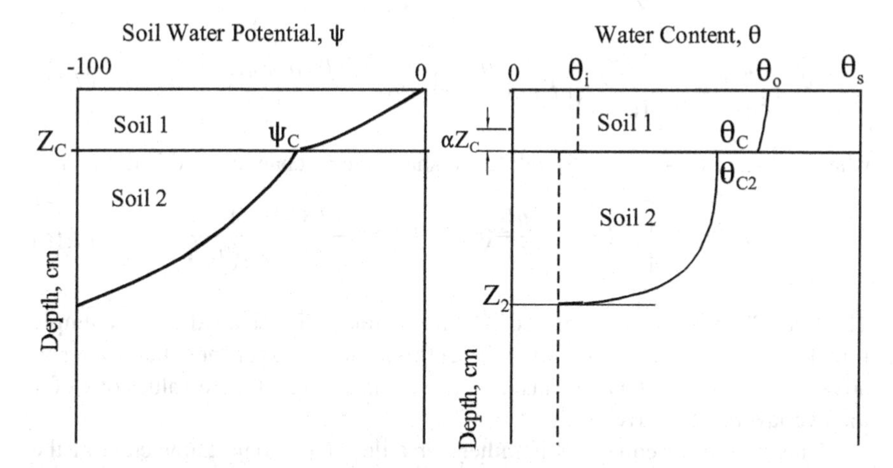

Figure 7.6. Definition sketch for the general two-layer soil profile for which the Corradini-Smith model is developed. There is a common value of ψ at the layer interface, but there is a discontinuity in θ.

A continuity equation like Equation (7.5) is written for both the surface and the interface. In the surface layer, the water total is

$$I_1 = Z_b\left[\alpha(\theta_o - \theta_{1i}) + (1-\alpha)(\theta_{1b} - \theta_{1i})\right]$$

and then a volume balance equation for the surface layer is

$$v_o - v_{12} = \frac{dI_1}{dt} = Z_b\left[\alpha\frac{d\theta_o}{dt} + (1-\alpha)\frac{d\theta_{1b}}{dt}\right] \quad (7.12)$$

In this equation the distribution of additions to the water content at the surface and by the interface is explicitly made, using the partitioning value α. The capacity variable $C_r(\psi)$ [see Equation (3.18)] expresses the slope of the retention relation, so that water content changes may be converted into capillary potential changes:

$$C_r(\psi) \equiv \frac{d\theta}{d\psi} \quad (7.13)$$

Then, combining these equations, a pair of ordinary differential equations are obtained for solution by the standard Runge-Kutta technique for two equations and two variables (ψ_o and ψ_b) [Corradini *et al.*, 2000]:

$$C_1(\psi_o)\frac{d\psi_o}{dt} = \frac{1}{\alpha Z_b}\left[v_o - K_{1b} - \frac{\phi_{1(b,o)}}{Z_b}\right] - \frac{1-\alpha}{\alpha}C_1(\psi_b)\frac{d\psi_b}{dt} \quad (7.14)$$

$$\frac{d\psi_b}{dt} = \frac{1}{P_L(\psi_b, t)}\left[K_{1b} + \frac{\phi_{1(b,o)}}{Z_b} - K_{2b} - \frac{\beta_2 p_2 \phi_{2(i,b)}}{I_2}\right] \quad (7.15)$$

where the term $P_L(\text{-})$ is mostly related to shape factor changes, and is defined as

$$P_L(\psi_c) = \left[\beta_2(\theta_{2c}) + \frac{d\beta_2}{d\theta_2}(\theta_{2c} - \theta_{2i})\right]\frac{I_2 C_2(\psi_c)}{(\theta_{2c} - \theta_{2i})\beta_2(\theta_{2c})} \quad (7.16)$$

The term $P_L(\text{-})$ is much simplified if, for example, β is assumed to be a simple function or a constant. Corradini *et al.* (2000) found by experiment that its derivative with respect to θ may be taken as a constant. C_1 and C_2 are values of C_r for the two layers, respectively.

This system of equations is rather versatile, given some knowledge of the soil hydraulic properties. Time to ponding can be estimated by taking $d\theta_o/dt = 0$, $\theta_o = \theta_{1s}$ and $v_o = r$. When $\theta_o = \theta_{1s}$ and $d\theta_o/dt = 0$ (ponded upper bound), one may solve for v_o representing an estimate of the system f_c.

Corradini *et al.* (1997) provided estimates of the relation of the factors βp to the flux v_o and a linear relation of β to θ_c and θ_i. Corradini *et al.* (2000) found, moreover, that with change of the partition coefficient α (or even keeping it constant) the model applied rather successfully to cases where the upper layer was restrictive (crust-layer type cases) and also to cases where the lower layer was restrictive; i.e., when the upper layer had higher hydraulic conductivity than the lower layer, and ponding occurs when the infiltrability limit $f_c(I_2)$ is reached.

Performance of the simplified model has been demonstrated with a variety of soil types and layer depths. Table 7.1 gives the hydraulic properties of some soils used here to demonstrate the ability of this simplification to model some complex situations. The infiltration model used in the simplification is the Green-Ampt model; there will be some limitation on the success with soils that do not correspond well to the delta-function diffusivity behavior.

Table 7.1. Hydraulic properties of soils used in demonstrating the simplified Corradini/Smith layer model.

Soil Name	K_s mm/h	θ_s	θ_r	ψ_B mm	ψ_a mm	λ	G mm	θ_i
1	0.4	0.3325	0.1225	-800	100	0.2	1104	0.1659
2	4.0	0.400	0.04	-400	50	0.4	476.5	0.0516
A	3.6	0.3325	0.1225	-800	100	0.2	1104	0.1659
B	4.0	0.400	0.04	-400	50	0.4	476.6	0.0516

Figure 7.7 illustrates the results of the simplified model in treating two soils (A and B in Table 7.1) that are absorption matched. This condition, suggested by Smith [1981], occurs when two soils have equal values of the product $K_sG\Delta\theta$, as in Equation (5.18). This means the two soils should behave the same under absorption or in the short time scale. Such soils do not however behave as one when they are layered together, as illustrated in this figure. The time of ponding is the same for either soil at the surface, however, if ponding occurs during the small time scale. The layer interface is reached soon after ponding, and the two soil profiles behave differently in this time period. At longer times, the two cases converge somewhat, but values of K_∞ will be different with different layering. Important here is the demonstration that the simplified model is able to respond relatively accurately to the complex deviations in the infiltrability patterns for either layering. It also reproduces the transitional surface and interface water contents. The approximate model simulates surface water contents nearly as well as Equations (5.50) or (5.51), despite the additional simplifications involved.

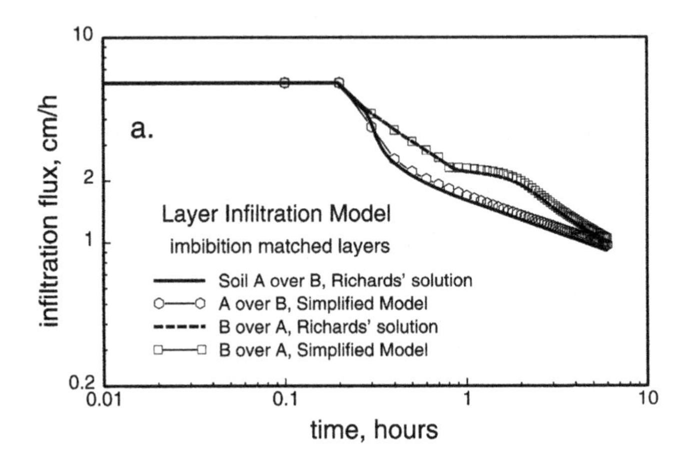

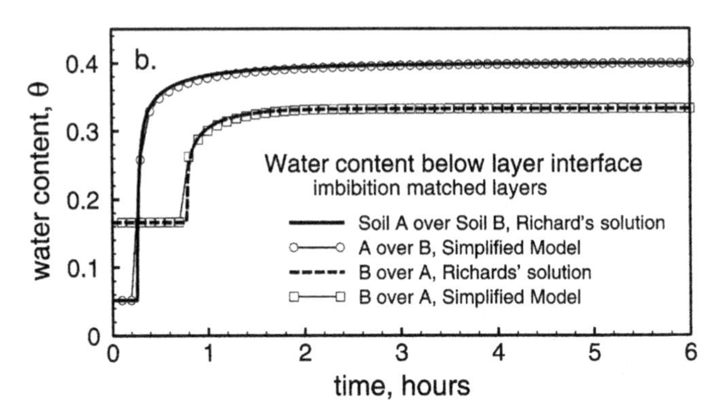

Figure 7.7. Comparison of the layered infiltration model of Corradini *et al.* with solution of the Richards' equation for the same case. The two soils used here are absorption matched, such that Equation (5.18) yields the same $f(I)$ for each soil since $G\Delta\theta K_s$ are equal. (a) infiltration pattern, and (b) θ below the interface.

These results also indicate an equally robust simulation by this simplified model of the rise in θ at the layer interface.

The results in Figure 7.8 represent a case where the more porous Soil 2 overlies Soil 1 [Table 7.1]. A uniform rainfall of 16 mm/h falls for 6 hours, and starts again after a hiatus of 10 hours. The example here is complicated by the fact that ponding occurs at the surface due to characteristics of the surface soil just before the advancing wetting front reaches the layer interface. At this point the infiltrability is further reduced, as shown in the upper part (a) of the figure. The approximate model follows each of these complexities rather well, with some bias in the simulation of the reduction in surface water content during the hiatus. The time

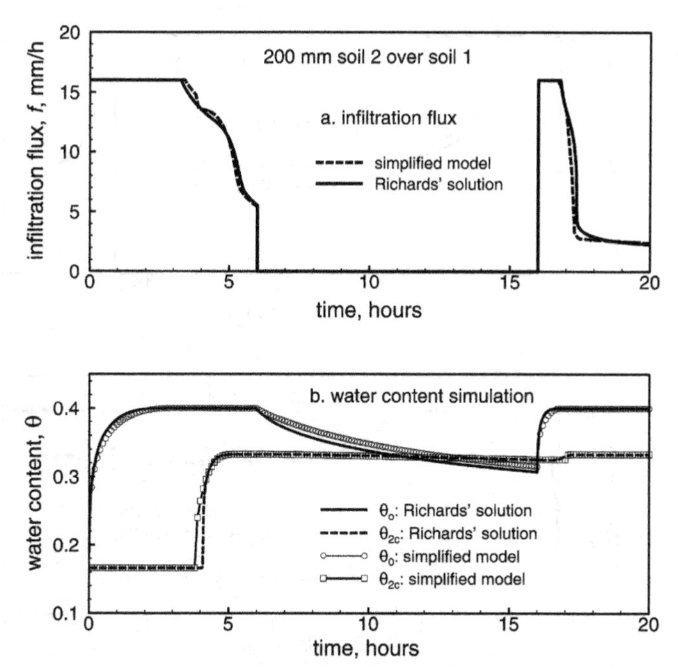

Figure 7.8. Simulation by the approximate Corradini-Smith model of a layered case where a more porous soil overlies a more restrictive soil. (a) the infiltration flux drops sharply after ponding when the lower layer is encountered, and (b) compares the approximate model simulation of the surface and interface θ with that from Richards' equation.

at which the wetting front reaches the lower layer is slightly underpredicted by the simplified model.

Figure 7.9 is a contrasting case to Figure 7.8, representing a restrictive 5 mm surface layer such as a crust. The rainfall used is the same as in Figure 7.8. For this case the simulation of the ponding time by the simplified model is not quite as accurate, but simulation of the reinfiltration infiltrability (upper graph) after 16 hours is quite good. While the redistribution of surface water during the hiatus is somewhat biased, with time it approaches rather than diverges from the more accurate Richards' solution.

SUMMARY

There are several simplified approaches to treatment of some of the complications of rainfall patterns and soil anisotropy with which one is confronted in field applications of infiltration theory. The methods retain the parameters used for the approximate integrations of Chapter 5, so that soil properties are represented by measurable parameters. However, some additional soil knowledge is needed if redistribution is to be estimated.

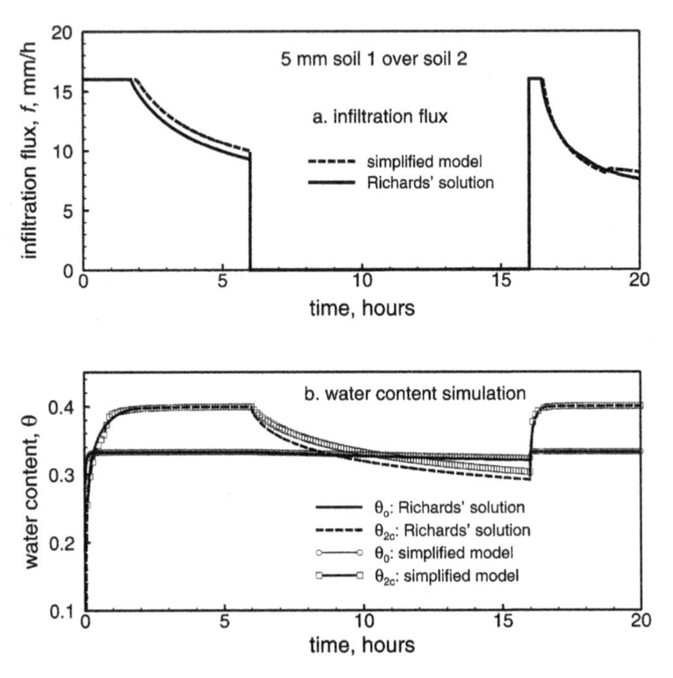

Figure 7.9. Simulation by the approximate Corradini-Smith model of a layered case where a more restrictive 5 mm deep soil (2) layer overlies the more porous soil (1). (a) the infiltration flux, and (b) comparison of the approximate model simulation of the surface and interface θ with that from Richards' equation.

Despite the fact that some reasonable values of the soil hydraulic characteristic parameters are required, there are hydrologic applications in which the reduced computational complexity of these simplified models is desirable. Most often, engineers are looking for an accurate simulation tool that does not require such knowledge or field measurements. It is not clear that good soil behavior models can be constructed without such information. But the necessity to perform a complete numerical simulation of Richards equation for relatively simple layering can be avoided with the approaches outlined here. In the next chapter, the field measurements that can be made to determine appropriate soil infiltration parameters are discussed.

8

Field Measurement of Infiltration Parameters

Keith R. J. Smettem

Soil Science and Plant Nutrition, The University of Western Australia, Australia

R. E. Smith

INTRODUCTION

In order to apply infiltration theory to practical problems, the appropriate infiltration parameters must be quantified. Of particular importance are the hydraulic conductivity and the soil's capillary properties, as defined by the sorptivity or capillary length scale. In order to make measurements that will reveal these parameters, simple or at least well-defined flow conditions must be created in the field in a way that appropriate measurements can be taken. Both these parameters cannot be found from a trace of flux or depth of infiltration at the small time scale, since only one parameter, sorptivity, is needed to characterise infiltrability in this region (Equations 3.13 and 3.14). Thus it is necessary to at least extend the time of observation until the medium time scale is reached. Table 8.1, based on representative soil texture class properties of Carsel and Parrish [1988], gives an indication of the values of some important soil scaling parameters that are important in designing and applying measurement methods. It should be emphasized that this table has the purpose of giving a general idea of the variations of these parameters, and any individual soil within these texture classes may have properties that differ significantly from the values shown. It should also be noted that texture class mean parameters of the Brooks and Cory hydraulic characteristics are given in Carsel and Parrish [1988], which, in the absence of other data, may be used in the redistribution models of the previous chapter.

Both agronomic practice and soil structural development exert a major influence on these infiltration parameters and it is therefore advantageous to make measurements *in situ*, with minimum disturbance of the soil surface. There are basically three main approaches to making simple, fast, and accurate measure-

Infiltration Theory for Hydrologic Applications
Water Resources Monograph 15

ments of infiltration behavior: sprinkler methods, ring infiltrometer methods, and permeameter methods. Each of these approaches is reviewed in this chapter, with particular emphasis placed on describing recent developments in the theory and practice of surface permeameter applications.

DRIP INFILTROMETERS

Drip infiltrometers, or rainfall sprinklers were originally developed by hydrologists and agricultural engineers interested in studying runoff and soil erosion by water. Mutchler and Hermsmeier [1965] have reviewed the various approaches taken in designing rainfall sprinklers to achieve an acceptable approximation of natural rainfall. Difficulties have been experienced in achieving a wide range of application rates while maintaining a drop size distribution and kinetic energy similar to that of natural rainfall. Trouble-free operation, portability and water requirement can also be of concern. Morin *et al.* [1967] describe a design with a rotating disc and Rawitz *et al.* [1972] describe the modified Purdue-Wisconsin sprinkler infiltrometer, which delivers low application rates with good aerial uniformity. This device does however, require a high frame tent, rotating shutter, pump and generator. Zeglin and White [1982] describe a simpler device, which uses compressed air and a solenoid valve to control the application rate but does not simulate natural drop sizes or energies. All these devices need to be transported using a small truck or trailer and use large quantities of water.

Table 8.1 Scale Parameters for Major Texture Classes, Dry Soil, from Carsel and Parrish [1988]

Texture Type Class	Typical K_s mm/h	Typical G mm	Typical dry soil S $mm/h^{1/2}$	Time Scale t_c, hr
Sand	30.0	82	38	0.80
Loamy Sand	15.0	97	29	2.0
Sandy Loam	4.4	165	21	11.
Loam	10.0	385	48	12.
Silt	2.5	914	37	109
Silt Loam	4.5	724	44	48
Sandy Clay Loam	13.0	240	43	5.5
Clay Loam	2.6	804	35	92
Silty Clay Loam	0.7	1590	26	680
Sandy Clay	1.2	589	21	73
Silty Clay	0.4	3570	29	2600
Clay	4.0	2230	73	167

Loch *et al.* [2001] introduced a rainfall simulator that uses Veejet 80100 nozzles mounted on a manifold, with the nozzles controlled to sweep over a plot of 1.5 m width. The frequency of the sweep controls the applied rainfall intensity and achieves high spatial uniformity. The device is portable and can be used on steep slopes. Loch *et al.* [2001] report that a field team of 3-4 experienced staff could run 2-3 plots per day.

Most sprinkler devices are set up to measure infiltration as the difference between applied rainfall and runoff from an experimental plot. Typically, the plot is bounded and runoff is routed through a small weir at the downslope end of the plot. The rainfall should be applied to an area larger than the bounded plot so that the plot samples what should be essentially vertical flow. A logging tipping bucket system can provide an accurate measure of runoff [Loch *et al.*, 1998]. The effect of surface detention storage on time to runoff remains unknown with this measurement procedure.

Rainfall sprinklers have generally been developed primarily for studies of soil erosion and so simulation of drop size and kinetic energy has been of importance. For studies interested only in infiltration, low energy sprinklers can be used. Ross and Bridge, [1985] describe one such device that can achieve low rainfall intensities (<10 mm/h) with high areal uniformity. Their device is highly portable and provides water to a target area of 1 m^2.

For unbounded plots, the time to incipient ponding [Rubin, 1966], rather than time to runoff is measured. Two approaches to observation of incipient ponding have been used. Bridge and Ross [1985] excavated a series of small shallow depressions within the experimental plot under the sprinkler and observed the time at which free water films developed in these depressions. White *et al.* [1989] and Smettem and Ross [1992] measured the time required to develop of a free water film over 50% of the plot area. A useful approach for measuring both the time to ponding and its distribution over the plot is to place small cans at one or more locations, and when free water films form in the area of the can, the time and the depth of water in the can is recorded.

Figure 8.1 shows typical experimental results of time to incipient ponding reported by Smettem and Ross [1992] for a fine sandy loam with a very dry antecedent moisture content of 0.04 m^3m^{-3}. The results compared well with the time to incipient ponding, t_p, predicted from the quasi-analytic solution Equation (6.5) given by White *et al.* [1989]. and Smith and Parlange [1978]

Also shown in Figure 8.1 are the 1 in 2 year and 1 in 50 year intensities for 1 hr duration rainfall events. The curves show that for dry antecedent conditions, only extreme rainfall events would generate runoff at this site.

It is evident that in ponding time Equation (6.5) the key soil hydraulic properties are G or S and K_s. It is possible to estimate S from rainfall sprinkler data when $r>>K_s$ by rearranging Equation (6.5) [Bridge and Ross, 1985]. However, it is now more usual to find rainfall sprinklers used to check predictions based on independently measured hydraulic properties. Permeameters are widely used to obtain these measurements and are therefore reviewed in detail below.

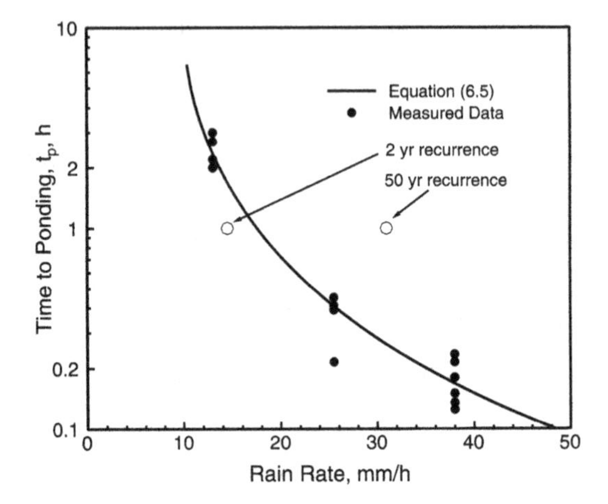

Figure 8.1. Time to incipient ponding at three rainfall intensities on a fine sandy loam with an initial water content θ_i of 0.04. Solid line is Equation (6.5).

RING INFILTROMETERS

This method has been in use for a long time, and requires inserting a confining ring or rings into the surface, and creating a shallow ponded condition in the ring. In early applications the water was added by hand and the depth changes measured with a point guage. A better method is to use a Mariotte tube which controls the surface depth and directly measures the water used.

There are several problems with this measurement method. Disturbance of the soil around the ring boundary is one concern. Sealant clay powders, applied at the boundary, have been used to minimize the effects of boundary gaps. If one wishes to create one-dimensional vertical flow, either the confining ring must be pushed very deep into the soil, or an outer ring should be used. In the case of a single ring, one-dimensional flow cannot last longer than required for the wetting to extend to the depth of insertion. This will rarely be long enough to achieve intermediate time scales that will allow fitting for estimation of K_s. The use of an outer buffer ring will allow the inner ring measurements to sample the inner part of a larger area, but that area also eventually becomes multidimensional, and the degree to which the inner ring flow approximates one-dimensional flow is demonstrated in Figure 8.2. The results shown are from a two-dimensional numerical simulation of a single and double ring inserted 10cm into a soil whose properties are given below in Table 8.3 (soil no. 2). The quality of approximation to one-dimensionality will be a function of the soil properties, especially the time scale [Wu *et al.*, 1997]. The time scale value t_s for this soil is approximately 48 minutes. The 10cm insertion depth is unusually deep, but restrains flow to one-

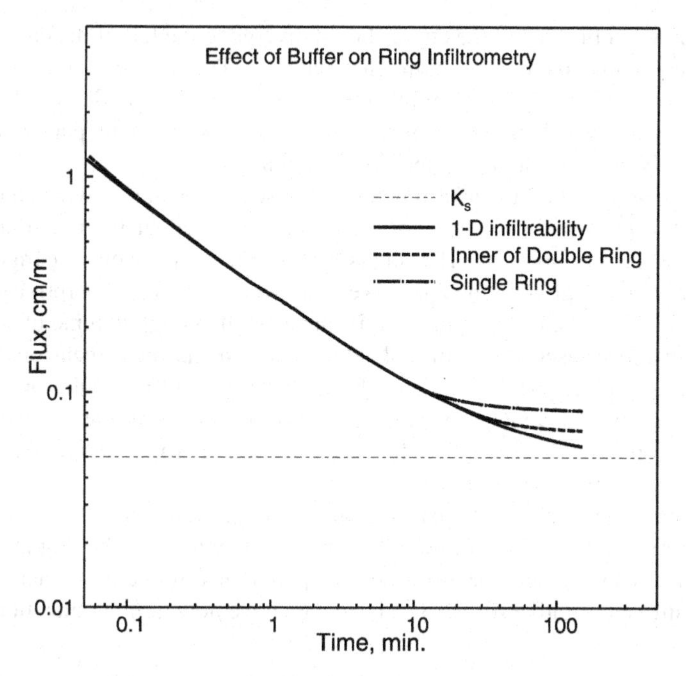

Figure 8.2. Mathematical simulation of flow through a double ring infiltrometer demonstrates that one-dimensional flow is not preserved in the inner ring, but the result is a compromise between a single ring and one-dimensional flow.

dimensional for only about 5 minutes. For the next 5 minutes, the flow is difficult to distinguish from one-dimensional, but after 10 minutes, the flow diverges significantly from one-dimensional, and is asymptotic to a value f_∞ that is larger than K_s for either the buffered or unbuffered ring cases. Clearly a very large buffer ring can constrain inner ring flow to nearly one-dimensional, even if its size would usually be quite impractical. However, the intermediate time scale would not be reached in a reasonable time such as150 minutes for many of the soil texture types shown in Table 8.1. The time scale limitation applies equally to the permeameter applications discussed below.

As an illustrative application of the Green-Ampt model to ring infiltrometer data, we employ measurements from a USGS report [Hofmann *et al.*, 2000]. In this case the outer buffer ring was very large, 3.5m in diameter, and contained three small inner rings of 0.75m diameter. Data was recorded with a pressure transducer. The outer buffer ring was inserted 15cm into the soil, confining the flow to one-dimensional for about 2 hours. Figure 8.3 illustrates the graphical analysis of this data using Equation (6.15). Time is plotted as the ordinate since this equation has $t(I)$ rather than $I(t)$. The data from ring 1 are much more uniform and give more confidence in the fitted values of $G\Delta\theta$ and K_s. Based on the results shown in Figure 8.2 the curve should diverge to the left of the data slightly at the commencement of multidimensional flow. Thus the data at larger times should not be given much weight in fitting Equation (6.15). With the variability of the data from ring 2, not uncommonly encountered in the field, parameter values cannot be very accurately determined from this ring.

One variation on the ring infiltrometer is the use of a small ring which has been termed a "sorptimeter" [Smith, 1999]. The ring is only about 10cm. in diameter, which allows easier insertion. The ring is designed to measure one accurate point in the short-time region on the $\boldsymbol{I}$(t) curve to allow calculation of sorptivity using Equation (3.13). The only assumption is that a small enough volume of water is used compared to insertion depth so that the flow remains one-dimensional in the short time solution space. This depends again on the soil hydraulic properties. One significant disadvantage of ponded infiltrometers is that the positive head boundary conditions can amplify the effect of macroporosity in comparison to the behavior of the surface during rainfall.

The appropriate analysis of data from single ring infiltrometers with a Mariotte tube requires the application of multidimensional flow theory. The ring infiltrometer can be used with the same theory for interpretation as for permeameters, which allow controlled negative surface heads, to which we now turn our attention.

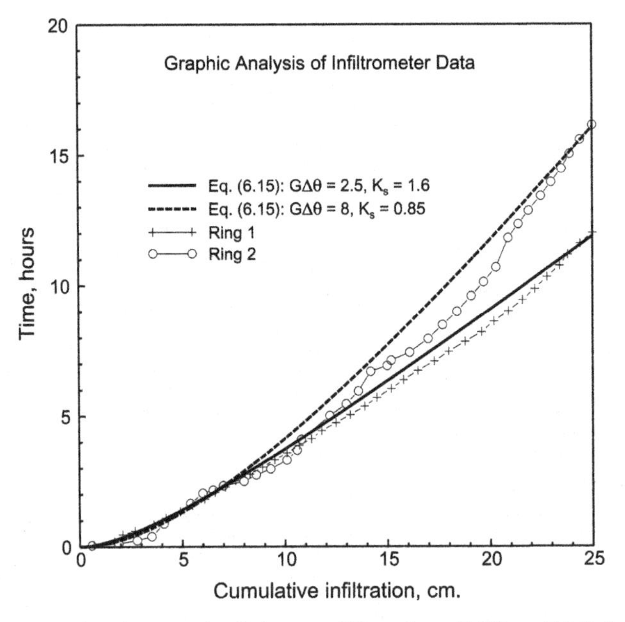

Figure 8.3. Double ring data may be fit by use of Equations (6.12) or (6.21), but the determination of K_S is difficult if the time scale of the soil is greater than the time for the advancing wetting front to exceed the confines of the rings. This example shows typical scatter of data, and the determination of parameters is limited to one or two significant figures.

PERMEAMETERS

The past decade has seen continuing development of the design and analysis of data obtained from permeameters. A permeameter may be distinguished from an infiltrometer by its ability to control within a limited range the pressure head at the soil surface during infiltration [White *et al.*, 1992].

The major advantage of permeameters is that they are portable and use relatively small volumes of water. This makes them particularly useful for studies of spatial variability [Smettem, 1987; Mohanty *et al*; 1998], or for characterizing the infiltration behavior of soils over watersheds.

Determining Infiltration Parameters by Monitoring Flow from a Shallow Circular Pond

The assumptions for soils behind the development of multidimensional flow theory for the analysis of permeameter applications are the same as for one-dimensional infiltration theory. It is assumed that the soil is rigid and isothermal, uniform over the depth of wetting and has a constant initial water content. What is often unstated is that the soil must not exhibit preferential (non-uniform) wetting. This is a point to which we shall return later.

The theoretical underpinning for measurement of infiltration parameters based on multidimensional flow devices was initiated by Philip [1966] when he explored solutions to the flow equation

$$\frac{\partial \theta}{\partial t}=\frac{\partial}{\partial x}\left(D\frac{\partial \theta}{\partial x}\right)+\frac{\partial}{\partial y}\left(D\frac{\partial}{\partial y}\right)+\frac{\partial}{\partial z}\left(D\frac{\partial}{\partial z}\right)-\left(\frac{\partial K}{\partial z}\right) \tag{8.1}$$

where z is the vertical ordinate in the cartesian coordinate scheme (x,y,z).

By employing his earlier series approach, Philip [1966] provided solutions to the axially and spherically symmetric forms of Equation (8.1). Of particular interest here is the steady flow v_∞ that is finally achieved. For a circular surface source of radius R, $v_\infty = q_\infty / \pi R^2$ where q_∞ is the steady-state volume flux (L^3T^{-1}).

Philip gave exact absorption solutions (neglecting the gravity term $\partial K/\partial z$ in Equation (8.2) for two extreme examples that provide an envelope to possible soil behaviour: the case where D is constant for all θ and the other extreme given by the Dirac δ-function diffusivity.

Philip [1966] also stated that the exact solutions did not apply when the effect of gravity was included in (8.1) for infiltration. Later, Philip [1968] presented solutions for steady infiltration from buried point sources and Wooding [1968] presented solutions for a shallow circular pond on the soil surface. Both solutions made use of the linearizing exponential hydraulic conductivity function of Gardner [1958] (Equation 3.30):

$$K(h) = K_s \exp(\alpha h) \; ; \; h_n < h < 0 \quad (3.30)$$

where h is the soil water pressure head and h_n is the initial soil water pressure head.

Clothier *et al.* [1995] point out that Philip [1966] had noted the utility of Equation (3.30) for deriving steady flow behaviour in two- and three- dimensional systems and that Philip had also commented on the highly practical implication that v_∞ does not depend on K_s alone but also depends strongly on the capillary properties of the soil.

Drawing on these foundations, Wooding [1968] considered the problem of flow from a shallow circular pond and developed a quite complex series solution for the steady flow v_∞ from the pond. Wooding [1968] did however, make a remarkable observation that the series solution could be approximated by a straight line:

$$v_\infty = K_S\left(1 + \frac{4}{\pi R_O \alpha}\right) \quad (8.2)$$

where R_o is the pond radius.

Although Weir [1987] added a refinement to describe the enhanced flow that occurs when capillarity dominates for the condition $\alpha R_o/2 < 0.4$, it is Equation (8.2) that has provided the theoretical underpinning for developing practical methods for estimation of infiltration parameters from circular surface sources. However, the difference between Wooding's and Weir's solutions is only about 10%, and this is substantially less than the variation usually encountered in the field [Smettem and Clothier, 1989].

Equation (8.2) can be expressed in dimensionless terms,

$$v^*_\infty = 1 + 4/(\pi R^*) \quad (8.3)$$

where $v^*_\infty = v_\infty / (\pi R^2 K_s)$ is the dimensionless steady-state flow rate and $R^* = R\alpha$ is the dimensionless pond radius.

From Equation (8.2) it is evident that for large values of R* (>10), the total steady-state flux is only about 10% greater than that due to gravity alone. However, for small values of R* (say, <0.15) the total steady-state flux is an order of magnitude larger because of the dominant effect of capillarity.

Equation (8.2) can also be expressed as

$$v_\infty = K_S\left(\frac{4\phi}{\pi R_o}\right) \quad (8.4)$$

where ϕ ($L^2 T^{-1}$) is the Kirchhoff transform, or 'matric-flux potential' (Equation 2.7) defined by [*Gardner*, 1958]

$$\phi = \int_{h_n}^{0} K(h)dh \quad -\infty \leq h_n \tag{2.7}$$

Substituting Equation (3.30) into (2.7) produces

$$\alpha = (K_s - K_i) / \phi \tag{8.5}$$

In addition, note that, considering the definition of G (or λ_s of Chapter 4) in Equation (5.14), we have $\alpha = 1/G = 1/\lambda_s$. In field soils drier than the so-called field capacity, K_i is usually very small relative to K_s [Scotter *et al.*, 1982] and can be neglected, thereby simplifying Equation (8.5) to

$$\alpha \approx (K_s) / \phi \tag{8.6}$$

The relation between ϕ or G and S was discussed in Chapter 5, and in terms of flux potential, can be given as

$$\phi = \frac{bS^2}{\theta_s - \theta_i} \tag{8.7}$$

The parameter b is a shape factor that was introduced in Chapter 5, and is constrained between the limits of ½ (Dirac δ-function soil) and $\pi/4$ (constant $D(\theta)$ soil). Because field soils rarely conform to these extremes, a reasonable intermediate value of 0.55 is often assumed [White and Sully, 1987; Smettem and Clothier, 1989].

Armed with Equations (8.2, 8.4 and 8.7) it was evident to many workers that it was technically possible to estimate soil infiltration parameters by measuring the steady-state flow from a circular surface source. To achieve this goal the capillarity and gravity terms contributing to v_∞ must be separated.

Scotter *et al.* [1982] proposed that the separation of the capillarity and gravity terms could be achieved by measuring v_∞ for two surface rings of widely different diameters. Equation (8.4) can then be solved by simultaneous equations to yield K_s and ϕ.

For rings of radius R_1 and R_2 we have

$$K_s = (v_{1\infty} R_1 - v_{2\infty} R_2) / (R_1 - R_2) \tag{8.8}$$

and

$$\phi = b[(v_{1\infty} - v_{2\infty}) / (R_1^{-1} - R_2^{-1})] \tag{8.9}$$

The sorptivity then follows from Equation (8.7), provided the water content difference is also measured. A cautionary note is that the minimum ring radius should be $>2\alpha^{-1}$ [Weir, 1986].

Before examining some of the practical difficulties that can be encountered with the twin ring approach it must be recognised that there is no 'standard' method of characterising soil hydraulic properties for all field conditions. Ultimately, the practitioner must decide which method is most appropriate for a particular set of circumstances. Comments on the limitations of a particular methodology are therefore provided as a guide.

The first practical concern is that Wooding's [1968] solution is strictly only valid if the depth of ring insertion into the soil does not influence the three-dimensional character of the wetting profile. Secondly, the ponding depth in the ring must be shallow enough to exclude the influence of hydrostatic pressure-head effects on infiltration.

Reynolds and Elrick [1990] have presented a solution that specifically accounts for these effects. Their motivation was also to provide an analysis for large ponded heads (up to 1 m or more) in order to increase flow rates in low permeability materials. Their resulting infiltration equation is

$$q_\infty = \frac{R_o}{W}(K_s H + \phi) + \pi R_o{}^2 K_s) \tag{8.10}$$

where H (L) is the steady depth of ponding in the ring and W is a dimensionless 'shape factor' coefficient that is determined using numerical simulations based on realistic flow geometry and the Richards equation for three-dimensional saturated-unsaturated flow. The W values thus obtained take into account the complex interactions between ring radius, R_0, depth of ring insertion, d, depth of ponding in the ring, soil capillarity and gravity [Reynolds and Elrick, 1990]. Note that Equation (8.10) reduces to Wooding's equation for the special case where $H = d = 0$ and $W = 0.25$.

Reynolds and Elrick [1990] provide tables of W values for different scenarios. For a range of representative soil hydraulic properties, depth of ring insertion between 0.03 and 0.05 m, ring radii, R, from 0.05 to 0.1 m and depth of ponding from 0.05 to 0.25 m the following useful relation was established between W and d/R_0.

$$W = 0.316(d/R_0) + 0.184 \tag{8.11}$$

Equation (8.11) greatly simplifies the determination of W for practical applications and provides increased flexibility in the choice of ring radius and depth of ring insertion.

The effect of field variability on parameter estimation must also be considered and therefore, a number of replicate measurements must be made to characterise a particular site. Scotter *et al.* [1982] recognised that K_s is usually log-normally distributed and therefore used geometric mean values of v_∞ for each ring size in their calculations. However, Smettem and Collis-George [1985] have shown that if the soil contains a random areal distribution of predominantly vertically orientated cylindrical macropores (with orifices open to the supply surface) the skew of the measured v_∞ distribution increases with decreasing ring radius.

It must also be noted that if the wetting pattern becomes highly irregular due to rapid and chaotic movement of water through macropores that are open to the surface water pond, the resulting infiltration parameters have no precise physical meaning. Furthermore, parameters measured under such conditions are of little use for predicting rainfall infiltration in the field up to the time of ponding. White *et al.* [1992] refer to their spectacular lack of success in attempting to undertake such predictions using hydraulic properties obtained by ponded techniques: 'It dawned on us finally that ponded measurements are heavily influenced by preferential (flow) pathways, which do not participate in rainfall infiltration until the soil-water potential at the soil surface approaches zero' [White *et al.*, 1992]. This observation, originally noted by Clothier and White [1981] provided the motivation for renewed development of surface disk measurement devices that could generate a negative soil water pressure head at the supply surface and therefore exclude the effects of soil macropores.

Surface Disk Tension Infiltrometers: Steady-State Solutions

Modern technical development of the single disk tension infiltrometer can be attributed to Ian White and colleagues at the Australian CSIRO Environmental Mechanics Laboratory during the early 1980's. The history of this development is documented by White *et al.* [1992], and the design is reported by Perroux and White [1988].

White [1988] proposed a mixed transient and steady-state approach for obtaining infiltration parameters from tension infiltrometer data. By introducing Equation (8.7) into (8.4) and rearranging to solve for K_0, White [1988] gave

$$K_o = v_\infty - \frac{4bS_o^{\,2}}{(\theta_o - \theta_n)\pi R_o} \tag{8.12}$$

The subscript o now denotes conditions for the negative soil water pressure head at the supply surface.

White [1988] assumed that the initial transient flow out of the tension infiltrometer was essentially one-dimensional and that capillarity effects were so dominant over gravity that the gravity free relation of cumulative infiltration I to S of Equation (3.13) holds: $I = S_0 t^{1/2}$

From Equation (3.13) it follows that the sorptivity can be estimated from

$$S_o = dI/dt^{1/2} \tag{8.13}$$

The question of the time period over which Equation (8.13) is unaffected by either geometry or gravity effects poses a problem for the application of Equation

(8.12). Sorptivity will be overestimated if either or both effects are present. This error then carries over into estimation of the hydraulic conductivity using Equation (8.15), with the magnitude of the error in the calculated hydraulic conductivity approximately proportional to the square of the error in the sorptivity measurement. The approximate time scales for the applicability of the gravity-free relation for various soil types are given in Table 8.1.

A further problem is that estimation of sorptivity using Equation (8.13) can be achieved over a much smaller wetted soil volume than applies to calculating the hydraulic conductivity at quasi-steady-state using Equation (8.12). The assumption of uniform initial conditions and uniform soil hydraulic properties over the depth of wetting may be violated by using a solution that requires measurements over two different wetted volumes. If any of the above sources of error are present it is not uncommon to obtain negative values of K_0 from Equation (8.12) [Hussen and Warrick, 1993; Logsdon and Jaynes, 1993].

Smettem and Clothier [1989] noted some of the difficulties that can arise by using a measure of sorptivity from early time infiltration to obtain K_0 from Equation (8.12). Following from Scotter *et al.* [1982], they employed multiple disk radii to measure v_∞ and then solved (8.12) by regressing v_∞ against $1/R_0$. The intercept of this regression gives K_0 and the slope is $4bS_0^2/\pi(\theta_0\text{-}\theta_i)$. This method removes the need to measure infiltration at early times but still requires measurements to be made at different locations. Because large macropores are excluded from the flow, tension infiltrometer data usually have much lower coefficients of variation than ponded ring data [Smettem, 1987; White *et al.*, 1992]. However, negative intercept values (K_0) have been reported [Logsdon and Jaynes, 1993], indicating that an unacceptably high sensitivity to local field variability is introduced by the requirement to average v_∞ from a number of measurements at each R_0.

To remove the noise in $K(h)$ estimates caused by spatial variability, a number of researchers have presented solutions that are based on imposing a sequence of wetting pressure heads with a single disk [Ankeney *et al.*, 1991; Reynolds and Elrick, 1991; Logsdon and Jaynes, 1993].

Ankeney *et al.* [1991] measured the quasi-steady state infiltration rate, q_∞, (L^3T^{-1}) at two (or more) negative wetting pressure heads, h_1 and h_2. Wooding's equation then gives

$$q_\infty = \pi R_o^2 K(h_1) + 4R_o\phi(h_1) \tag{8.14}$$

$$q_\infty = \pi R_o^2 K(h_2) + 4R_o\phi(h_2) \tag{8.15}$$

Ankeney *et al.* [1991] derived an approximate expression for the difference $\phi(h_1)$ - $\phi(h_2)$ and assumed a constant ratio between K and ϕ. This gives three equations with three unknowns, which can be solved simultaneously for the two conductivities $K(h_1)$ and $K(h_2)$. Sequential measurement of q_∞, at progressively more negative wetting pressure heads can be undertaken quite rapidly in the field but the effect of drainage-induced hysteresis is unknown [White *et al.* 1992].

Reynolds and Elrick [1993] also based their solution on Wooding's equation but assumed that $K(h)$ was given by Equation (3.30). Introducing Equations (3.30) and (8.6) into Wooding's solution expressed in terms of q_∞ at saturation gives

$$q_\infty = (R_o/0.25\alpha + \pi R_o^2)\, K_s \exp(\alpha h_o) \tag{8.16}$$

The logarithmic transform of Equation (8.16) is

$$\ln q_\infty = \alpha h_o + \ln[(R_o/0.25\alpha + \pi R_o^2)K_s] \tag{8.17}$$

Equation (8.17) gives the linear relation between $\ln q_\infty$ and h_o. The slope of this relation gives

$$\alpha = \ln(q_1/q_2)/(h_1 - h_2) \tag{8.18}$$

As in the Ankeney *et al.* [1991] method, the subscripts 1 and 2 refer to sequentially more negative pressure heads imposed at the supply surface. K_s is then obtained from the intercept on the $\ln q_\infty$ axis (y-axis) of the plot vs h_o on the x-axis. The expression for K_s is

$$K_s = \frac{0.25\alpha q_1}{R_o(1 + 0.25\alpha\pi R_o)(q_1/q_2)^p} \tag{8.19}$$

where $p = h_1/(h_1 - h_2)$

$K(h)$ is then obtained from Equation (3.30).

If α is observed to change with h_o (by visual inspection of the $\ln q_\infty$ vs h_o plot) then the procedure can be implemented by piecewise fitting to each sequential pair of $q_\infty(h_o)$ measurements.

A number of researchers have reported that in field soils the slope of the $\ln q_\infty$ vs h_o plot changes dramatically at about –30mm to –40mm h_o [Clothier and Smettem, 1990; Smettem and Ross, 1992; Jarvis and Messing, 1995; Mohanty *et al.* 1997)]. Logsdon and Jaynes (1993) also recognised this observation by limiting their non-linear regression technique to wetting pressure heads of –30 mm or less.

These steady-state methods of analysis have been widely used over the last decade and comparisons have been reported by Logsdon and Jaynes [1993] and Cook and Broeren, [1994]. Perhaps the major limitation to consider before employing any quasi-steady-state analysis is the assumption that a quasi-steady-state has actually been achieved during an infiltration test. The time to quasi-steady-state increases with increasing ring radius and if not reached, v_∞ will be overestimated.

It is also important that the wetted volume does not encounter significant non-uniformities within the soil profile over the period of the infiltration test.

Surface Disk Tension Infiltrometers: Transient Solutions

The potential practical limitations to solutions based on Wooding's equation have led to an interest in developing simple transient flow solutions for the tension disk infiltrometer [Warrick, 1992; Smettem *et al.*, 1994; Haverkamp *et al.*, 1994]. Due to the shorter experimental time, a smaller wetted volume of soil can be sampled under transient flow conditions and therefore, conditions of homogeneity are more likely to be encountered. Transient infiltration tests can also be performed over shorter experimental times than quasi-steady-state tests and are therefore particularly appealing for studies of spatial variability.

Smettem *et al.* [1995] have proposed that a practical experimental time limit, t_{exp}, may be defined by a profile depth, z_1, over which hydraulic properties can be considered uniform. Their estimate, based on an expression for the wetted profile depth not corrected for gravity, is

$$t_{exp} = \frac{z_1(\theta_o - \theta_n)^2}{S_o^2} \tag{8.20}$$

The experimenter must specify z_1 from practical experience, but in agricultural soils the depth of tillage is a useful guide.

Turner and Parlange [1974] presented an approximate analytical expression for the lateral flux at the periphery of a horizontal one-dimensional infiltration process under transient axisymmetric conditions. Smettem *et al.* [1994] showed that the expression of Turner and Parlange [1974] could be modified to account for the geometric effects of a surface disc source. For imbibition, the side effect due to the axisymmetric flow geometry is linear with time:

$$I_{3D} - I_{1D} = \frac{\gamma_D S_o^2}{R_o(\theta_o - \theta_n)t} \tag{8.21}$$

where the subscripts 3D and 1D denote three-dimensional and one-dimensional infiltration and γ_D is a proportionality constant that for normal working conditions lies within the range $0.6 < \gamma_D < 0.8$ [Haverkamp *et al.* 1994]. For imbibition, I_{1D} is given by Equation (3.13).

This can be compared to the short-time solution given by Chu *et al.* [1975] which is relevant for two-dimensional edge effects in linear heat diffusion. Using a constant diffusion coefficient $D^* = \pi S_o^2/4(\theta_o - \theta_n)^2$, an equivalent flux expression was developed by Warrick [1992]

$$v \cong 0.5\, S_o\, t^{-1/2} + 0.885\, S_o (D^*)^{1/2}/R_o \tag{8.22}$$

The integration of Equation (8.22) with respect to time gives

$$I_{3D} = S_o t^{1/2} + \frac{0.885\, b^{1/2} S_o^{\,2} t}{R_o(\theta_o - \theta_n)^2} \tag{8.23}$$

Comparing this expression to Equation (8.21) gives $\gamma_D = 0.784$, which is within the bounds given by Haverkamp *et al.* [1994] and similar to the optimal value of 0.75 obtained from the experimental results of Smettem *et al.* [1994].

Haverkamp *et al.* [1994] introduced Equation (3.13) into (8.21) and included a gravity term to give a physically-based three-dimensional infiltration equation that is appropriate for short to medium times:

$$I_{3D} = S_o t^{1/2} + \left[\frac{2-\beta}{3} K_o + \frac{\gamma S_o}{R_o(\theta_o - \theta_n)}\right] t \tag{8.24}$$

where β_H is an integral shape constant constrained to $0 < \beta_H < 1$ [Haverkamp *et al.* 1994].

Equation (8.24) can be simplified to a two-term infiltration equation of the form

$$I_{3D} = C_1\, t^{1/2} + C_2 t \tag{8.25}$$

where $C_1 = S_o$ and the term C_2 contains both the lateral capillarity flow and gravity flow terms in Equation (8.24). The same physical interpretation applies to the solution of Warrick [1992] if the coefficient C_2 describes the second term in (8.23). Zhang [1997] also uses a similar two-term infiltration equation but only relates C_2 to gravity forces, which is questionable and not consistent with Wooding's [1968] equation, as pointed out by Vandervaere *et al.* [2000a].

The flux expression from the time derivative of Equation (8.25) is simply

$$v_o = 0.5\, S_o t^{-1/2} + C_2 \tag{8.26}$$

Vandervaere *et al.* [2000a] give an insightful discourse on different fitting techniques to obtain C_1 and C_2 from Equation (8.25). They point out that although it is a well posed nonlinear least squares optimization problem it is in fact, ill-conditioned because of intercompensation between $t^{1/2}$ and t. In consequence, the estimated values of C_1 and C_2 can be highly dependent on the chosen optimization technique. They recommend use of a 'differentiated linearization' (DL) technique, used previously to examine the hydraulic properties of crusted soils [Vandarvaere *et al.* 1997]. Performing this differentiation on Equation (8.25) gives

$$\frac{dI_{3D}}{dt^{1/2}} = C_1 + 2C_2 t^{1/2} \tag{8.27}$$

where $dI_{3D}/dt^{1/2}$ is approximated by

$$\frac{dI_{3D}}{dt^{1/2}} = \frac{\Delta I}{\Delta t^{1/2}} = \frac{I_{i+1} - I_i}{t_{i+1}^{1/2} - t_i^{1/2}} \quad (i = 1..n-1) \tag{8.28}$$

in which n is the number of data points, and the corresponding $t^{1/2}$ is calculated as the geometric mean

$$t^{1/2} = [\sqrt{t_i t_{i+1}}]^{1/2} \quad (i = 1..\ n-1) \tag{8.29}$$

Vandervaere *et al.* [2000a] note that the plot of $\Delta I/\Delta t^{1/2}$ vs $t^{1/2}$ should be linear, with C_1 equal to the intercept and C_2 equal to half the slope. If this is not the case, then the fitted values are likely to have no physical meaning.

Vandervaere *et al.* [2000b] point out that the precision with which K_o can be estimated will depend on the relative importance of gravity and lateral capillarity on the total flux from the permeameter. The permeameter infiltration measurement is situated in the gravity domain under the condition:

$$\frac{\gamma C_1^2}{R_o(\theta_o - \theta_n)} < \frac{C_2}{2} \tag{8.30}$$

In Equation (8.30) K_o is the dominant term in C_2 and can be reliably estimated by introducing C_1 and C_2 into Equation (8.24) and rearranging to give

$$K_o = \frac{3}{2-\beta}\left[C_2 - \frac{\gamma C_1^2}{R_o(\theta_o - \theta_n)}\right] \tag{8.31}$$

K_o is however a minor term in C_2 if the permeameter infiltration measurement lies within the lateral capillarity domain given by

$$\frac{\gamma C_1^2}{R_o(\theta_o - \theta_n)} > \frac{C_2}{2} \tag{8.32}$$

If Equation (8.32) applies, the operation is badly conditioned and estimation of K_o is unreliable. The experimenter must therefore try to satisfy the conditions of Equation (8.30) and may be able to achieve this by judicious choice of initial conditions and disk radius. Because sorptivity is highly dependent on initial water content, C_1 will diminish markedly with increasing initial water content, whereas C_2 should remain quite constant if the material is rigid.

It is, however, still important to note that because the lateral component of the flow from the disk infiltrometer stabilises the infiltration flux, the source geometry cannot be neglected when determining sorptivity at short times and C_1 should therefore be determined from Equation (8.27).

A Field Example of Tension Infiltrometer Data Analysis

Tension infiltrometer data at a supply pressure of –40 mm was obtained from a tillage trial on a fine sandy loam in South Australia. Cumulative infiltration readings were performed manually and a thin (5-6 mm) sand cap was used to obtain good contact between the disk and the soil. This is a common field procedure to create a flat contact surface without disturbing the surface soil. Measurements were performed post-harvest on minimum tilled plots under very dry antecedent conditions. Initial and final water contents were obtained with TDR.

Figure 8.4 illustrates an application of the differential linearization (DL) method to one particular replicate measurement. Note the early time 'noise' in the data due to the confounding effect of the sand cap and pressure equalisation as contact is established between the disk and the capping material. This noise is not obvious in the either the $I(t)$ or $I(t^{1/2})$ curves. Resolving exactly which part of the early-time $I(t^{1/2})$ curve to use if the 1-d approximation is applied to obtain S is particularly problematic in this case. A comparison of the best estimates of S with the 1-d 'early-time' approximation and the DL method analyses are provided in Table 8.2. In all but one case, the DL method returns much lower values of S than the 1-d approximation.

Table 8.2. Comparison of sorptivity estimates from tension infiltrometer field measurements, with supply potential of –40mm.

Replicate	Sorptivity Estimate, 1-D 'Early Time'	($mm\ h^{-1/2}$) DL Method
1	12.1	9.4
2	20.9	11.4
3	20.2	5.6
4	92	16.1
5	20.7	5.9
6	26.1	14.1
7	12.4	11.9

It is important to note the role that the sand cap may play in altering the position of the flux curve. The sand should have two properties in comparison to the underlying soil: it needs to have a saturated conductivity that is greater than the underlying soil, and it needs to have a value of ψ_B higher than the applied tension at the disk interface. The latter requirement is important so that the conductivity in the capping sand does not drop below that of the soil during unsaturated flow, and thus turn the sand into a restrictive surface layer. These two relative property conditions are not achievable for all soils, capping materials and applied heads.

Figure 8.5 shows the cumulative infiltration data for replicate 2, together with curves constructed using the DL derived parameters. In this particular case, the

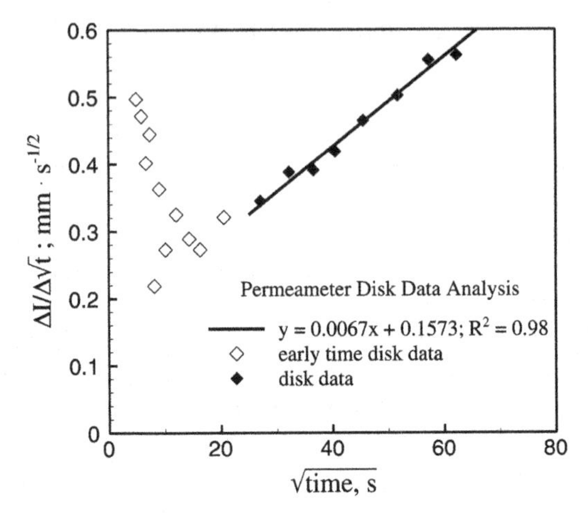

Figure 8.4. Illustrative example of the DL method for hydraulic parameter estimation from tension infiltrometer data.

effect of the sand contact layer can be compensated for by modifying Equation (8.25) to give (Vandevaere *et al.*, 2000)

$$I_{3D} = I_s + C_1 t^{1/2} + C_2 t \tag{8.33}$$

in which I_s is the water holding capacity of the sand layer. Also shown is a curve constructed by identifying C_1 with 'early time' (1-d approximation) sorptivity. The overestimation is now clearly evident in the cumulative curve.

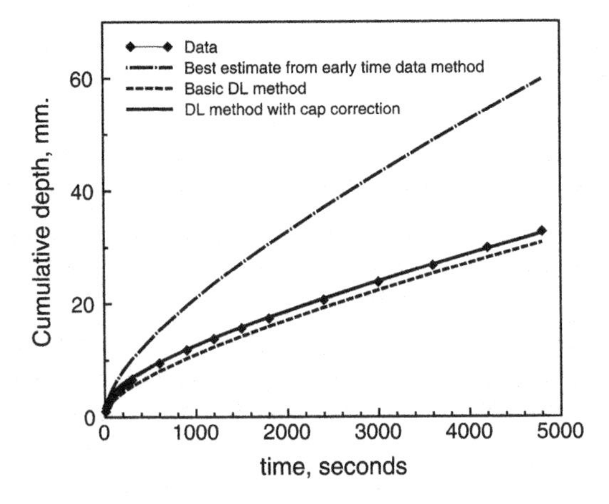

Figure 8.5. Cumulative infiltration data and the infiltrability curves found by several methods of data analysis.

Surface Disk Tension Infiltrometers: Numerical Investigation of an Illustrative Layered Profile

Although Smettem *et al.* [1995] proposed a practical time limit over which hydraulic properties might be considered uniform during wetting, there has been no detailed investigation of layering on surface measured hydraulic properties. An illustrative example is presented here using results from simulations performed with a slightly modified version of the SWMS2D model [Simunek, 1994]. Infiltration from a 20 cm diameter disk into a 'uniform' and 'layered' soil profile was investigated. Soil properties were obtained from Smettem and Ross [1992]. The uniform profile comprised a fine sandy loam. For the layered profile a denser layer was introduced at 10 cm depth to typify the properties of the plow pan that is often encountered in this soil type. Simulations were performed over 150 min. using soil water pressure heads of -100 mm and -20 mm H_2O. Relevant soil hydraulic properties used in the simulations are presented in Table 8.3.

Simulation results, Figure 8.6 and 8.7, illustrate that for this particular case at least, the presence of the layer has little effect up to 100 minutes on the quasi-steady-state disk infiltration rates for either the -100 mm or -20 mm supply pressure heads. For both simulated examples the disk flux is not equal to one-dimensional flow, even at small times. The early part (up to ~20 min) of the disk flux curves are identical for the layered and unlayered cases, and thus the interpretation to obtain sorptivity [Equation (8.26)] is unaffected. However, the time scale for this soil is relatively short, and many soils would not approach multidimensional quasi-steady flow in the few hours usually taken for measurement. For each application head, a one-dimensional vertical flux is affected by the layer more severely and at a much earlier time than disk flux, as shown in these two figures.

The higher application head, Figure 8.7, demonstrates more of the effect of gravity compared with the lower (-100mm) head, Figure 8.6, which has a longer period of time in which the flux is quasi-absorption in behaviour. It is evident that the flux for this case is approaching a steady flow after about 100 mins and even after 30 mins the flux is only 10% higher than the value at 150 mins. For the -20 mm supply pressure head the final flux at 150 mins is 2.7 times higher than for the -100 mm supply head and again, at 30 mins it is only 10% higher than the value at 150 mins.

Table 8.3 Soil hydraulic parameters for simulated disk infiltrometer. Values are defined with respect to Equation 2.20

Soil	K_s mm/h	θ_s	θ_r	ψ_B mm	c	λ
2 =loose loam	30.0	0.400	0.10	100.	3.0	0.20
3 = compact loam	6.0	0.400	0.10	250.	3.0	0.20

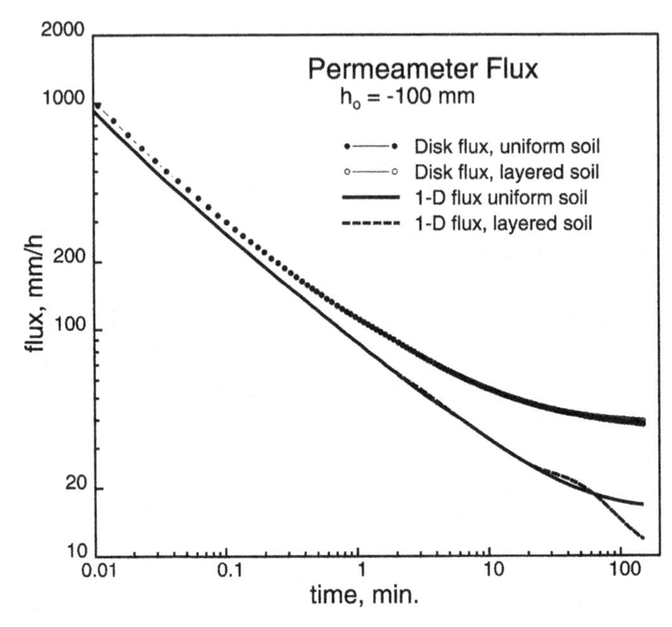

Figure 8.6. Infiltration fluxes from simulated tension infiltrometer: comparison of uniform and layered profile results at boundary head -100mm.

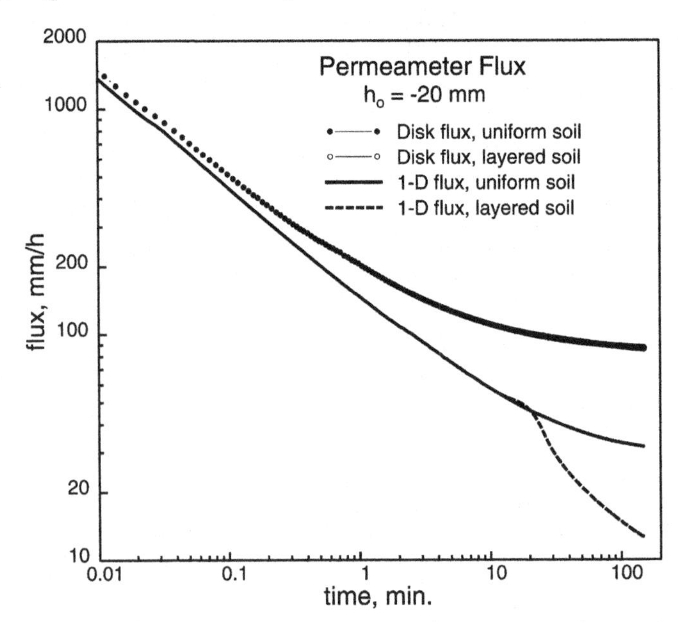

Figure 8.7. The same permeameter conditions as in Figure 8.6, but with the surface head increased to -20mm. In both cases the one dimensional flow is more sensitive to the soil layer than is the permeameter result.

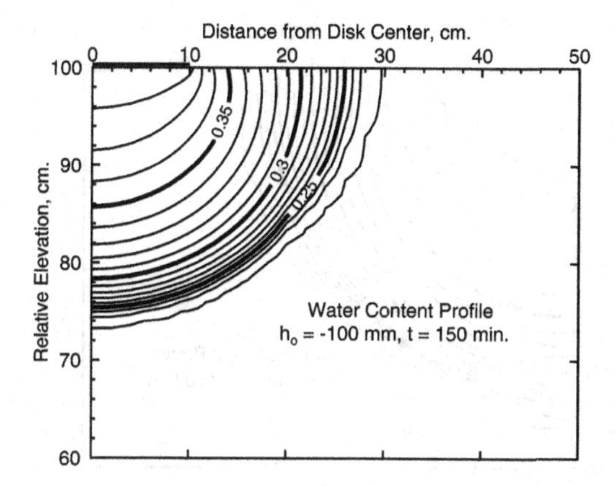

Figure 8.8. Water content profile under disk for uniform soil type 2 (Table 8.3) at 150min under imposed head of -100mm.

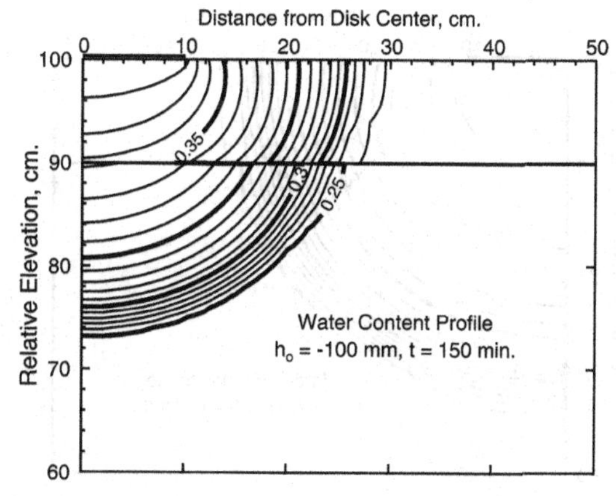

Figure 8.9. Water content profile under disk for layered soil with 10cm of soil type 2 over soil type 3, under imposed head of -100mm.

Wetting water content profiles at 150 min. are shown in Figures 8.8 - 8.11. The profiles show a slightly more pronounced lateral spread in the top 10 cm of the layered profiles for both supply pressure heads. The perturbation due to the layer is clearly seen in Figures 8.9 and 8.11. If the contrast between the layer properties was more extreme then this perturbation and lateral spread in the top layer would be expected to increase. Nevertheless, it is encouraging for the use of solutions based on Wooding's equation that the layer change simulated here did not significantly affect the final steady-state flux.

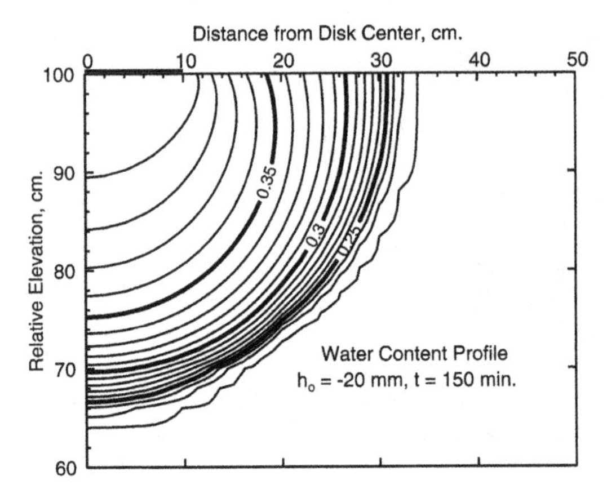

Figure 8.10. Water content profile under disk for uniform soil type 2 (Table 8.1) at 150min under imposed head of -20mm.

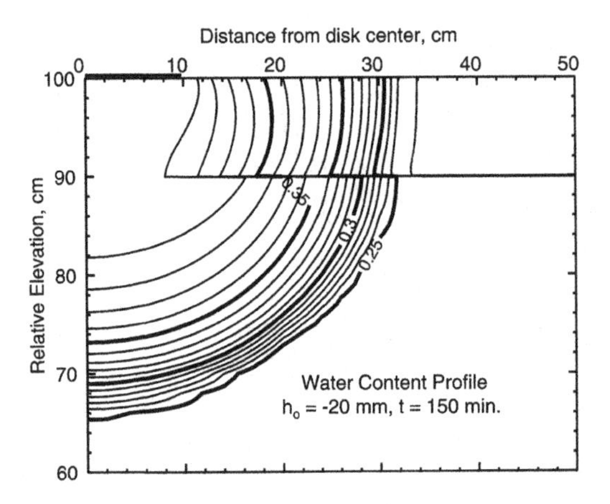

Figure 8.11. Water content profile under disk for layered soil with 10cm of soil type 2 over soil type 3, under imposed head of -20mm.

CONCLUDING REMARKS

The use of rainfall sprinklers and permeameters to measure infiltration parameters has been discussed. Particular attention has been paid to the theoretical development of solutions to the problem of flow from circular surface disk sources. The advantages and problems associated with quasi-steady-state and transient analyses have been identified. It may be concluded that there is no 'universal' procedure for measuring infiltration parameters in all soils. The transient method can be used to identify the sorptivity using the accurate solution to Equation (8.27) but problems can arise when estimating the hydraulic conductivity if flow is in the capillarity time domain. Results from the numerical simulations suggest that Wooding's solution may be more robust than hitherto thought because profile non-uniformities may have to be relatively shallow and quite severe to affect the steady flux through the surface layer. However, the analyses reported here also indicate the dangers of effects of choice of capping sand on the measured results. Clearly, further investigation will be necessary to identify conditions under which profile layering affects the measured flow rates from surface disk sources.

9

Infiltration and Runoff on a Hillslope

D.A. Woolhiser, contributing co-author

Formerly with Agricultural Research Service, USDA
SW Watershed Res. Center, Tucson, AZ

INTRODUCTION

The traditional approach to modeling infiltration on a hillslope is illustrated in Figure 9.1. The temporal distribution of infiltration rates, $f(t)$, is subtracted from the temporal distribution of rainfall intensities $r(t)$ to create a temporal pattern of rainfall excess, $e_r(t)$:

$$e_r(t) = r(t) - f(t) \tag{9.1}$$

In this method, the rainfall excess and the infiltration losses last only as long as the rainfall exceeds $f(t)$, and the pattern $e_r(t)$ is routed across the watershed to the channel system.

The routing methods used in hydrologic models have been many and varied, including linear storages and simple translation in time [Crawford and Linsley, 1966; Freeze, 1980]. More realistic routing methods are more common today, usually based on surface water hydraulics such as the kinematic wave simplification of the de St. Venant equations [Singh, 1996]. In any case, shallow depths of water move slowly over the surface, and there is in fact an opportunity for infiltration as long as water is on the surface. In other words, $e_r(t)$ may be negative, because the source of infiltration flux $f(t)$ is not limited to the rainfall, but includes water at the surface created earlier when $e_r(t)$ was positive. The infiltration time may last much longer than the period of rainfall. For this reason at least, the interaction of runoff and infiltration is important, and in this chapter some of the key effects of this interaction on hydrologic response of complex hillslopes will be discussed. The effects discussed include those of surface microrelief, and various forms of spatial variability.

Infiltration Theory for Hydrologic Applications
Water Resources Monograph 15

DOI NUMBER

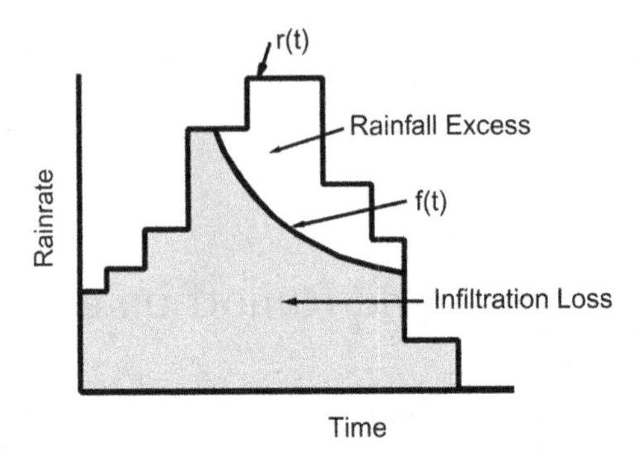

Figure 9.1. The traditional method of separating rainfall rate into rainfall excess and infiltration does not allow for the infiltration of water that exists on the surface after the end of rainfall.

GENERAL SURFACE FLOW AND INFILTRATION INTERACTION

In fact Equation (9.1) is correct only as long as $r(t) \geq f(t)$, i.e., while e_r is positive. The existence of a depth of surface water during surface runoff does not change the local validity of Equation (9.1) only as long as e_r is positive, but local water depth is a significant source of influx and a boundary condition determinant at any time that $e_r(t)$ becomes negative. In this case, infiltration at any point continues at the value of infiltrability until the local surface depth supply is exhausted. The exhaustion of local water depth results from both the lateral flow of surface water and from loss into the soil due to the continuation of f_c. This creates a complexity in the treatment of rainfall "losses" that has not always been recognized in hydrologic models. Not only is there a dependency in time once surface water has been created, but there is a spatial dependency across the watershed, since upper parts of the runoff path will dry earlier than those at the downslope portions of a watershed, where water depths will be replenished by flows from above. This interactive process was modeled about 30 years ago by Smith and Woolhiser [1971].

The effect of this interaction on the runoff hydrograph at the outlet of a simple uniform plane is shown in Figure 9.2. The hydrographs in this figure were created using a simple kinematic wave simulation of the surface water hydraulics. While this process is not treated here in detail [see Eagleson, 1970; or Singh, 1996, for example], the assumption is that simple one-dimensional surface water flow may be described by a combination of a basic one-dimensional continuity equation;

$$\frac{\partial h}{\partial t} + \frac{\partial q}{\partial x} = r(t) - f(t) \tag{9.2}$$

plus a nonlinear equation relating unit surface discharge, q, to effective local flow depth:

$$q = b_r h^m \tag{9.3}$$

in which h is local effective surface water depth, b_r is a coefficient dependent on slope and hydraulic roughness, and m is an exponent greater than one. Distance along the slope is designated by x. The solution of this equation may be obtained numerically, or analytically using the method of characteristics, as used for the kinematic approximation to soil water flow in Appendix I.

If e_r is calculated from the rainfall pattern alone, as is traditional in many hydrologic procedures, the hydrograph recession limb is excessively extended, as seen in Figure 9.2. Indeed, with no recession loss value [$e_r = 0$], the recession is theoretically asymptotic to $q = 0$ and extends indefinitely. The degree of error caused by neglecting recession infiltration is related to the relative infiltration rate at the time of cessation of rainfall. In this example, the infiltration rate is a significant fraction of the rainfall rate at the end of rainfall, so the loss rate to which the receding surface water flow is subjected is relatively high. This is not an uncommon occurrence where Horton runoff is encountered. For storms which are large and long and reduce the value of f_c to a relatively small value, the effect of recession infiltration would be less than illustrated in Figure 9.2. In general, proper interactive treatment of infiltration is necessary to correctly describe the hydrograph recession.

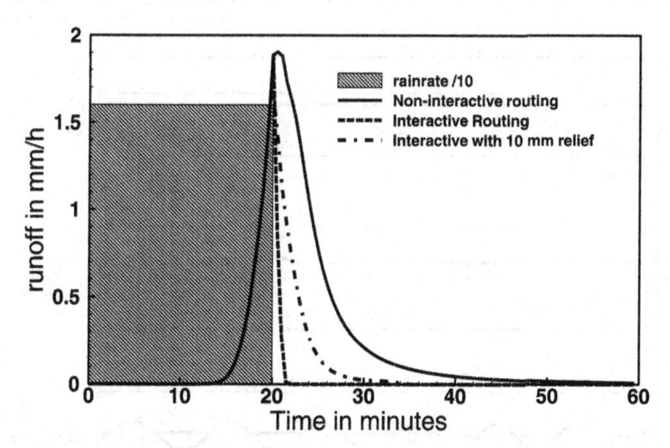

Figure 9.2 Effect on hydrograph recession of various methods of treating infiltration during recession, when rainfall has fallen below f_c. The solid line illustrates the traditional method where infiltration losses are subtracted only from the rainfall pattern. Treating infiltration interactively with a smooth surface results in the dotted recession hydrograph, and including the effect of surface micro-relief will again modify the recession as shown in the dot and dash curve.

EFFECT OF SURFACE RELIEF ON SURFACE WATER LOSSES

Natural catchment surfaces are not simple uniform plane surfaces as may be assumed by the above simple example. There are generally irregularities in the surface, which may consist of uniform or random micro-relief or a rill network into which flow is channeled when surface runoff occurs. Figure 9.3 illustrates this microrelief in terms of hypothetical or idealized sections across the flow path on a hillslope surface. Relatively uniform micro-relief is created in agricultural areas by tillage. Random micro-relief may be present in minimally-eroded natural areas due to soil or vegetation distributions. In either case, flow of surface water is rarely uniform in depth, and generally is concentrated on certain parts of the surface. Thus the area for infiltration after rainfall ceases may be severely restricted, and the rate of loss from the surface during recessions may be much less than if it were assumed to cover the entire soil surface (Figure 9.2). Making this situation more complex, the relative area covered should increase monotonically with the relative amount of surface water. Conceivably, there is an amount of surface water storage that would cover the entire surface. As the surface water storage approaches zero, the area covered would also approach zero. For simplicity, further discussion of the hydraulics of surface water will assume one-dimensional flow. Actual flow paths on a catchment generally consist of large numbers of meandering microchannels. The overall effect may nevertheless be assumed one dimensional, with overall convergence or divergence at larger scales based on the topographic controls.

The value of h used in Equation (9.2) is an effective mean surface water depth, which may be defined as the volume of water stored on the surface per unit area.

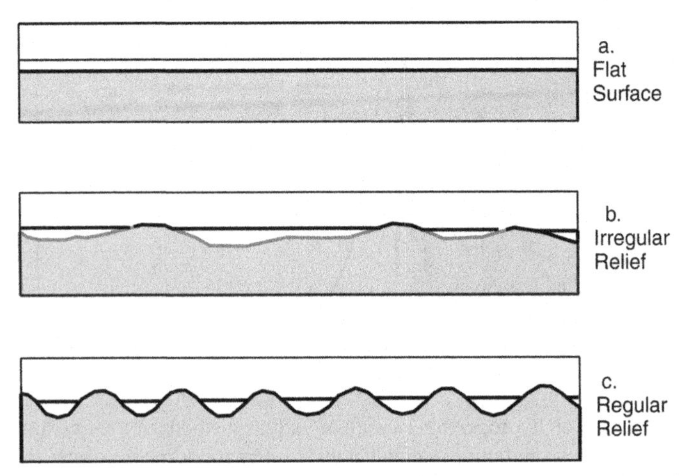

Figure 9.3. Schematic representation of 3 kinds of surface relief types. The flat surface (a) is an idealized condition seldom realized in nature, which is most often like that shown in (b). The regular variation shown in (c) is more like the surface often created by agricultural tillage.

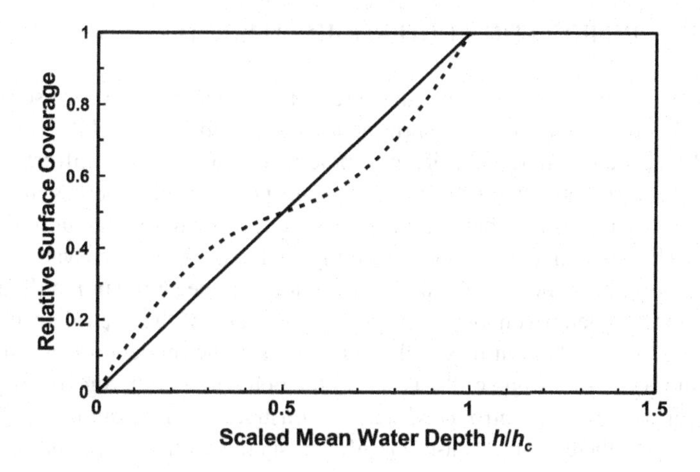

Figure 9.4. The effect of surface relief on the relative area covered by flowing water may be represented in scaled terms as shown. The dotted line might result from the relief pattern shown in Figure 9.3c, and might also be an approximation for the random pattern of Figure 9.3b.

This value will of course vary along the slope of the runoff path. In the nomenclature of open channel hydraulics, this value is the *hydraulic depth*; the cross-sectional area divided by the surface width. Note that in Equation (9.3) it is assumed that normal flow can be related to hydraulic depth. Conservation Equation (9.2) can easily be written in terms of hydraulic depth, as long as a relation $h(A)$ is definable.

Without microrelief, the value of h is equal to the actual water depth. Microrelief can be characterized in scaled variables as shown in Figure 9.4, relating the relative surface area covered to the effective mean surface water depth. In this relation the depth is scaled on h_c: the smallest value of h at which the surface is entirely covered with water. Note that the actual maximum relief, the elevation difference at a point along the surface between the lowest and highest point, is not the same as the value of h_c since h is found by dividing the flow cross- sectional area by the flow width. The surface flow may indeed have effective depths greater than h_c, but since the wetted area cannot increase, the curve must be flat for values of $h/h_c > 1$.

Various surface conditions will result in differences in this scaled curve, but a simple uniform variation suffices to demonstrate the effect of microrelief on recession flows. The hydrograph (c) in Figure 9.2 represented by the dot and dashed line results from assuming the linear microrelief relation shown in Figure 9.4, with maximum microrelief depth h_c of 10 mm. Increasing the value of h_c results in a more restricted area for infiltration during the period when infiltration is occurring from residual surface water alone, and thus the recession of curve (c) will approach more closely the solid curve, representing no infiltration after rainfall ceases.

INFILTRATION HETEROGENEITY AND RUNOFF

The complications of layering and intermittent rainfall patterns discussed in Chapter 7 are relevant to the determination of the infiltration flux pattern at a point. Perhaps even more significant to practical applications of infiltration theory are the variations in infiltration flux from point to point across an area of hydrologic significance. This is a major research challenge today in hydrologic science. The challenge includes describing and quantifying the variations, as well as determining the significance of variations on the simulation of hydrologic behavior of a heterogeneous area. This issue is one of the central problems in "upscaling" - the adaption of smaller-scale hydrologic models for use at larger scales. In this section some of the issues of the behavior of heterogeneously infiltrating surfaces will be introduced and demonstrated. While rainfall rates have been shown to have in some cases significant spatial variations [Goodrich et al., 1995], it is generally at a scale an order of magnitude larger than measured variations in soils. Thus the focus here will be on soil spatial heterogeneity and for simplicity the rainfall rates will be assumed uniform.

Spatial heterogeneity can take many forms. General and continuous changes in soil type across or up and down a runoff surface can be treated as deterministic variations. These may be associated with changes in the soil-forming process or depth, or associated in some other manner with geomorphologic features. A simple example of deterministic variation is simulated below. Other variations may be random or associated with apparently random variation in vegetation or vegetal clustering. This variation may also be associated with microtopographic heterogeneity. Because it is the most critical parameter, we are here focusing on variations in K_s. However, soil changes also result in spatial variations in parameters such as G and $\Delta\theta_{si}$ [see Chapter 5].

A heterogeneous runoff surface immediately poses the problem of spatial interactions. If each point on a runoff-producing area is different, and there is an interaction through the flow of water across the surface, there is opportunity for one point on the catchment to affect its neighbor. Runoff generated at an earlier time at an upslope location can provide an addition to the rainfall influx at a location, and thus significantly and suddenly change the infiltration at a point. This case has received some study in the literature [see Corradini, *et al.*, 1998, for example].

One method for treating infiltration heterogeneity is by the use of various kinds of *ensemble* simulation. This consists of simultaneous simulation of a variety of cases representing the range of variations, and using the combined result as the areal effective value. For example, Woolhiser and Goodrich [1988] represented a catchment as an ensemble of runoff flow strips, and simulated the sum of their responses as the catchment response. This is illustrated in Figure 9.5. Conceptually, a zero or first order catchment can be treated as composed of a number of flow strips of equal area, one of which is illustrated in this figure. Each strip can be assumed to exhibit (slightly convergent) one-dimensional flow.

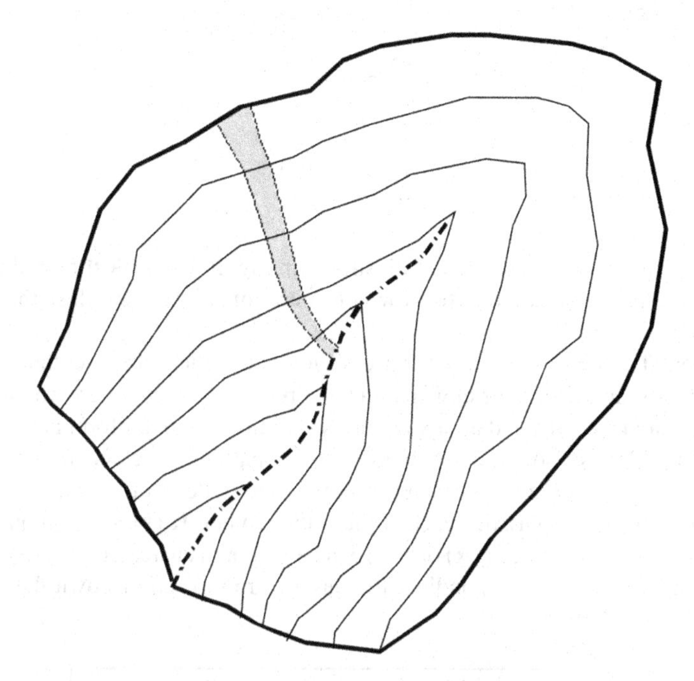

Figure 9.5. A simple order 1 catchment may be treated as composed of an ensemble of strips, one of which is shown here, and distribution of infiltration or other runoff properties may be treated by assigning different values to each strip.

This method implicitly assumes some lumping of soil conditions along the flow path, but retains the ability to include non-linear process variation. In general, for a non-linear system, there exists no mean value of a process parameter which, when used in the system, will duplicate the behavior of an ensemble of systems with distributed values of the parameter. Thus many of the studies reported below will use an ensemble model approach.

Deterministic Variation in Infiltration Rates

Woolhiser *et al.* [1996] studied the deterministic variation of K_s along a slope, in connection with the effects of the microrelief described above. The slope hydraulic response was assumed to follow the kinematic wave equations (9.2) and (9.3), and the infiltration model chosen was the Smith and Parlange model, Equation (5.41). To characterize the runoff behavior of a catchment with upslope or downslope trends in K_s, the kinematic wave equations were written and solved in characteristic form [Wooding, 1966]:

$$\frac{dx}{dt} = m\,b\,h^{m-1} \tag{9.4}$$

$$\frac{dh}{dt} = r - A_f\,f \tag{9.5}$$

Here, the factor A_f is the coefficient illustrated in Figure 9.4, reflecting reduction of areal effective loss rate due to flow concentration, and is only less than one while $r < f$.

The characteristic solution in x,t space (Figure 9.6) reflects the fact that during simple hydrograph rise, the runoff path is composed of a lower part where flow is uniform and unsteady, and an upper part where flow is nonuniform and steady. These two regions of solution are divided by the path of the characteristic from the upper boundary. Were e_r constant, runoff at the outlet would reach a steady value when this characteristic reached the outlet. When $K_s(x)$ varies monotonically along the slope, a trace $t_p(x)$ or $x_p(t)$ can be drawn in characteristic (x,t) space representing the advance of ponding, the onset of runoff, up or down the slope.

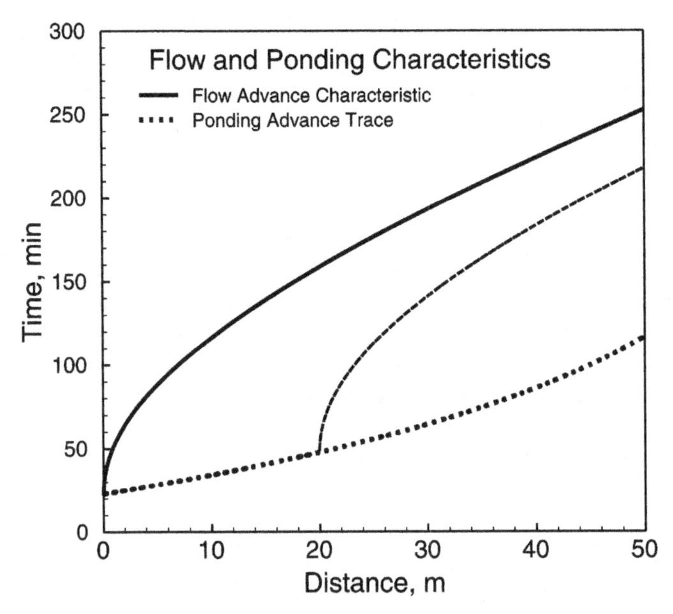

Figure 9.6. Characteristics of the kinematic description of overland flow begin at the time of ponding at each distance, x. The trace of the characteristic's starting point, $t_p(x)$, (dotted line) depends of any trends or variation of infiltration properties, such as K_s, along the slope. Here an increase in K_s with x is illustrated. Flow characteristics always start with zero velocity (vertical slope).

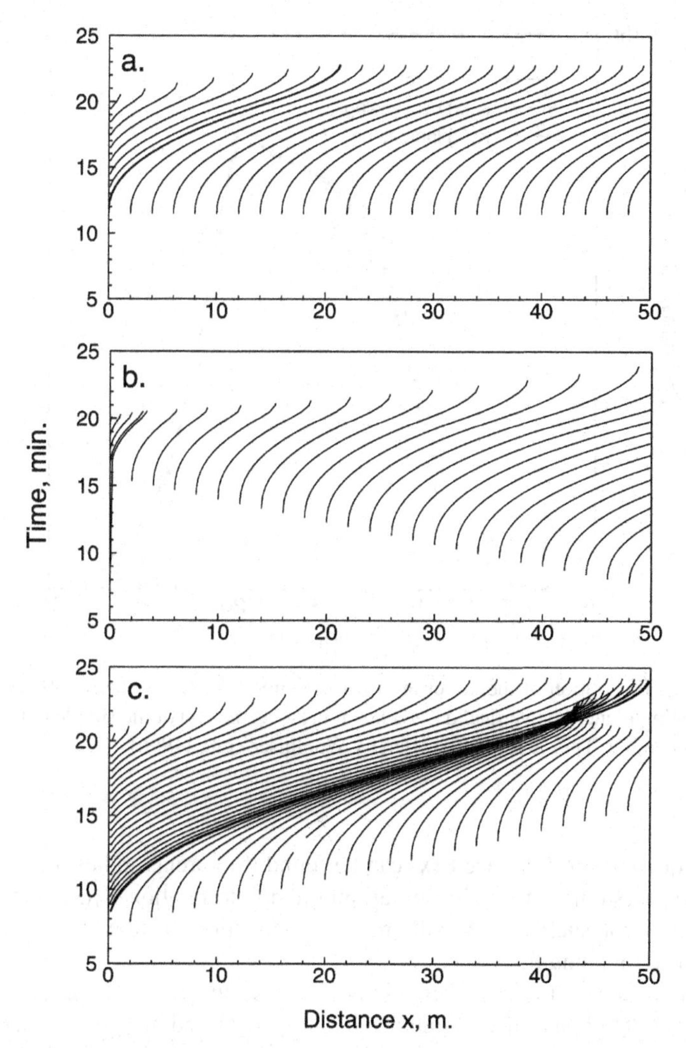

Figure 9.7. Characteristic curves for runoff on a simple plane surface having hydraulic conductivity (a) uniform, (b) decreasing with distance down the plane, and (c) increasing with distance down the plane.

Runoff characteristics begin at this trace, as illustrated in Figure 9.6. For K_s increasing in the downslope direction, runoff starts earlier at the upper end than the lower end, and vice-versa. Conversely, if $K_s(x)$ decreases downslope, the $t_p(x)$ moves upslope, and the runoff area increases upslope with time as for many cases of saturation-induced runoff. An illustration of characteristics originating all along the runoff slope for cases of uniform K_s, K_s decreasing, and K_s increasing, but with the same average K_s, is presented in Figures 9.7 a, b, and c, respectively.

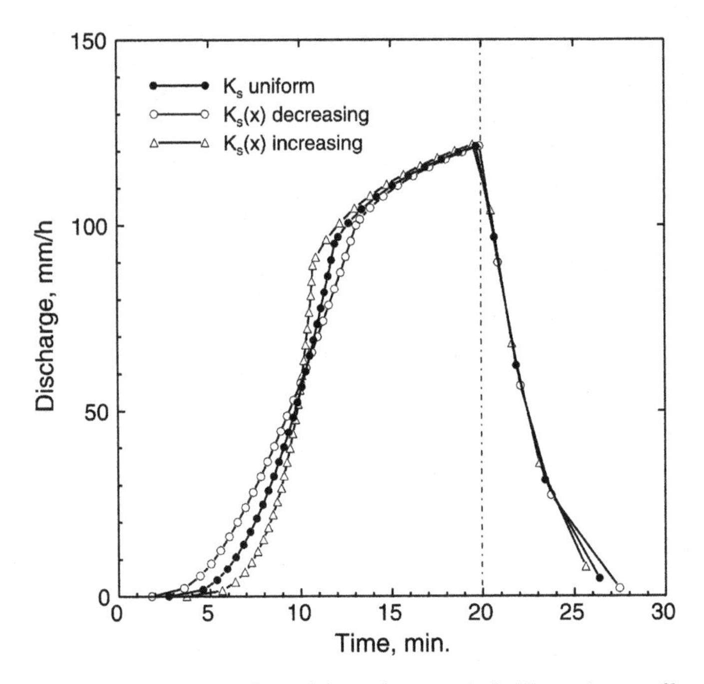

Figure 9..8. Illustration of the effect of downslope trends in Ks on the runoff response of an infiltrating plane. The vertical dashed line represents the end of the simple uniform rainfall of 177 mm/h. This case was reported by Woolhiser *et al.* (1996).

The equation for this trace $t_p(x)$ can be found if the relation between K_s and x is known. Assuming a steady rainfall and a simple relation $K_s(x)$, substitution into an equation such as (6.4) will produce an equation relating the ponding time to the location along the slope.

For large values of relative rainfall rate $r_* = r/K_s$, where the amount of runoff is a significant fraction of the rainfall depth, the spatial trend in K_s has a minor effect on the overall runoff amount, and results in simple changes in the hydrograph rise shape, as shown in Figure 9.8. Runoff starting earlier at the lower end of the slope results in a more slowly rising hydrograph, and runoff starting earlier at the top will produce a hydrograph that starts later but rises more quickly. This is a result of the nonlinear relation of Equation (9.3), with deeper flows traveling faster.

From Figure 9.8 it can be seen that were the rainfall excess period to be shorter following the ponding time, either by a reduction in the rainrate (causing a ponding later in the storm), or a cessation of rain during the hydrograph rise, there would be a significant difference in the runoff peak due to the trends in the value of K_s. This was demonstrated by Woolhiser *et al.* [1996] for slopes with the same average K_s, as shown here in Figure 9.9.

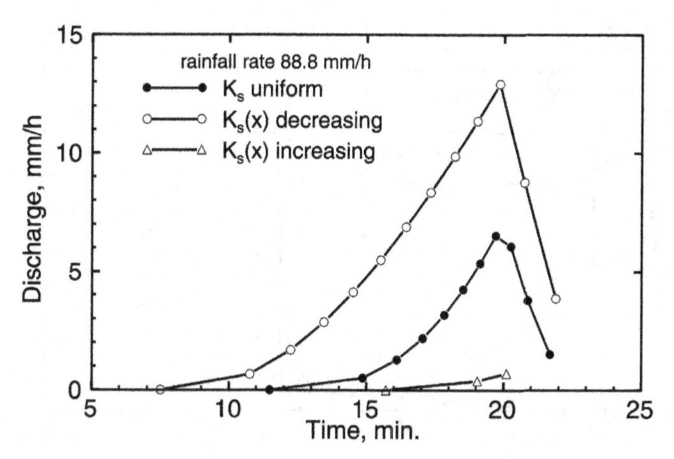

Figure 9.9. The effect of trends in Ks along the runoff path are far more significant for slower rainfall rates than for higher ones as shown in Figure 9.8.

One of the more dramatic aspects of the interaction of spatial trends in K_s and microrelief was demonstrated by Woolhiser *et al.* [1996] for the case of downslope increasing K_s. When runoff is generated upslope earlier than downslope, there is considerable surface storage that may infiltrate before reaching the outlet, provided rainfall ceases relatively early. When there is, in addition, microrelief that restricts the re-infiltration, this flow from the upper area may reach the outlet after rainfall has ceased. In addition, as shown by the characteristics in Figure 9.7c , this increasing flow or advance flow wave may steepen until a shock is formed (indicated by the intersection of characteristics). Figure 9.10 illustrates the hydrographs resulting from a downslope increasing K_s and three values of microrelief. The significance of the restriction from microrelief on recession infiltration opportunity is here apparent and dramatic. The shocks, illustrated by the vertical dotted lines, arrive at the catchment outlet after the rainfall has ceased. Moreover, without the trend in K_s and the existence of microrelief to restrict post- rainfall infiltration, the peak runoff rate would be negligible in comparison with the actual value. This case is discussed in more detail by Woolhiser *et al.* [1996].

Random Spatial Variation in Infiltration Rates

At smaller scales within a catchment area, there are in all cases random small scale variations in soil properties such as K_s. Spatial sample measurements of K_s have consistently found relatively significant variations, and the lognormal distribution is commonly a good description of the data from such sampling [Nielsen *et al.,* 1973; and Viera *et al.*, 1981, for example]. Several papers have appeared in the literature in which hydrologists have looked at the distribution of

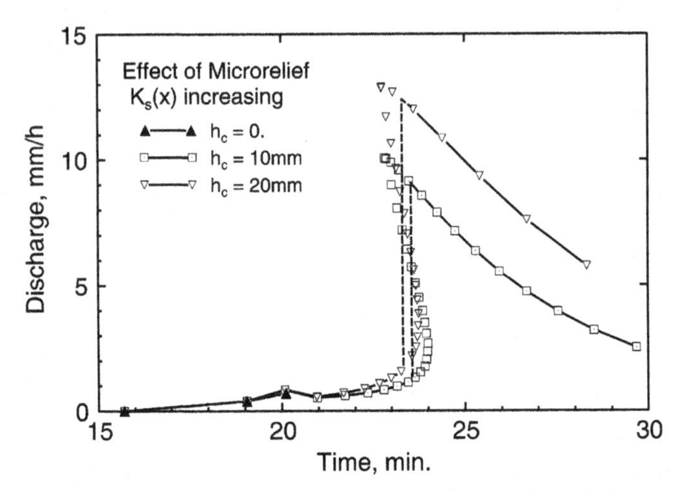

Figure 9.10. The combination of downslope increasing K_s and microrelief can cause a dramatic change in runoff peak, and a peak runoff that comes after the rainfall has ceased. The vertical lines indicate shock waves that advance down the plane when characteristics intersect, as in Figure 9.7c.

K_s and have taken several approaches to treating the effect of such variation on the infiltration and runoff of a catchment. [Sivapalan and Wood, 1986; Maller and Sharma, 1981; Smith and Hebbert, 1979]. In the following discussion the random distribution of K_s will be assumed to be lognormal with mean ξ_K and variance σ_K^2. The coefficient of variation is $CV_K = \sigma_K/\xi_K$.

Monte Carlo Sampling. Commonly, simulation of the effect of a randomly varying parameter in a system is done using Monte Carlo sampling, in which a random number generator is used with appropriate transformation so that a large number, N, of samples of a parameter of interest, in this case K_s, are generated with a desired statistical distribution. This set of parameter values may be used in a system to generate an output $f_n(t)$, (n = 1,N) for each sample parameter value, and these added so that an ensemble average behavior, $\bar{f}(t)$ is obtained. This is a random, large-sample method of simulating ensemble behavior, and has been done by Sharma, *et al.* [1980], and Smith and Hebbert, [1979], among others. The ensemble is assumed to be composed of the sum of the independent effects of each component added together.

Another approach to ensemble behavior was taken by Maller and Sharma [1981] and Sivapalan and Wood [1986], involving numerical or approximate integration of the first moment of the Philip two-term $f(t)$ relationship (Equation 5.28) with K_s expressed in terms of its probability distribution. Rather than random sampling, the infiltration function was used as a transform function between the probablility density of the parameter (K_s) and that of the infiltration rate. This involves either functional approximation and/or numerical methods.

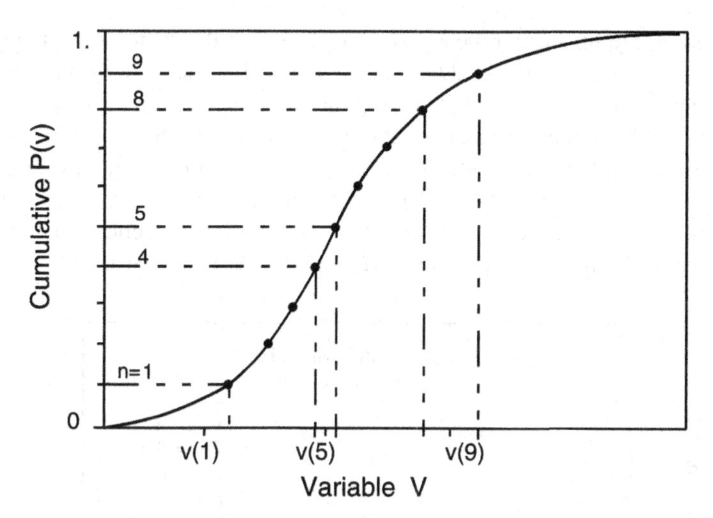

Figure 9.11. The Latin Hypercube method of treating random parameter variables is numerical rather than statistical, and divides the cumulative probability distribution of variable V into *M* portions of equal probability, as shown here for $M = 10$. Each portion *n* of the distribution is represented by the value, v(n), of its first moment.

Latin Hypercube Sampling Another more general method of numerical simulation of random parameter variation for any process, including the distributed infiltration process, is called the *Latin Hypercube* (*LH*) method. Figure 9.11 illustrates the method. It is a stratified sampling method that divides the probability density function into M equal areas, n = 1, ...M, and the centroid of each area is used to determine a sample value $K_s(n)$. Thus this method is a *numerical* simulation of a distributed parameter, as contrasted with the Monte Carlo method, which is a *random sampling* simulation. The major advantage of the *LH* method is that the parameter distribution may be simulated at least as accurately as with a Monte Carlo method while using an order of magnitude fewer sample values.

Ensemble Infiltration Behavior If one treats an area of soil as composed of an ensemble of points that act together without regard to spatial interaction, some form of sampling may be used to simulate an ensemble infiltration relation. With a distribution of an infiltration parameter such as K_s, there is a distribution of ponding times and the ensemble infiltration function is somewhat dispersed in time. Given that the distribution of ponding times is small compared with the time necessary for very shallow surface water to flow from one area to an adjacent soil area, this approach is defensible for simulation of distributed infiltrating areas under many conditions.

Several studies have looked at the ensemble infiltrability relation using Monte Carlo or numerical integration of the ensemble behavior. All have demonstrated that the ensemble infiltration behavior is 'diffused' in the region of ponding, as shown in Figures 9.12. The sudden onset of runoff associated with a ponding time (at which the value of f_c falls below the rainfall rate) for a single value of K_s, is replaced by a more gradual beginning of runoff. The amount of divergence from sudden runoff is a function of the degree of variation of K_s, as measured here by the value of CV_K, the coefficient of variation of K_s. This was demonstrated with Monte Carlo ensemble simulation by Smith and Hebbert [1979], and by approximate numerical means by Sivapalan and Wood [1986].

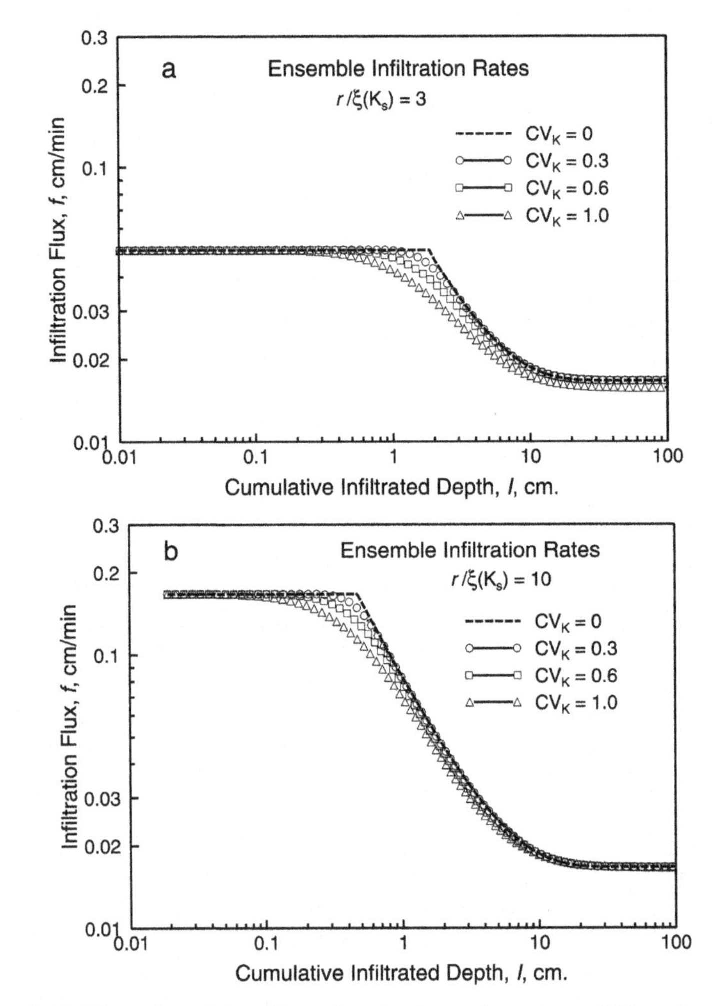

Figure 9.12. Illustration of the effect of randomly varying values of K_s on the ensemble infiltration dynamics. (a) with rainrate relatively small shows some bias in the the final asymptotic ensemble infiltration rate, which is not found at longer times when the rainrate is relatively high, as in (b).

Another effect of ensemble behavior with lognormally distributed values of K_s is a bias in the ensemble value of the large-time asymptotic value of f_c. At any point, as shown above, this asymptote is the local value of K_s. We will refer to the ensemble effective, large scale asymptotic value of infiltration rate as K_e. As pointed out several years ago by Hawkins and Cundy [1987] however, for asymptotic distributions of K_s [with a theoretical range from 0 to ∞] the effective value of K_e is dependent on rainfall rate, r, since for any value of r there is some portion of the area in which $K_s > r$. Mathematically, we represent a distributed K_s with a probability density function $p_K(\bullet)$, and cumulative distribution function $P_K(\bullet)$. The expected value of K_s is represented by $\xi(K_s)$ or ξ_K. The areal effective value of K_s, which we will here refer to as K_e, for a rainfall rate r, is not equal to ξ_K, but can be represented as

$$K_e = r\left[1 - P_K(r)\right] + \int_0^r p_K(k)\,dk \qquad (9.6)$$

The first term represents that part of the area, however small, for which K_s is greater or equal to r. The integral part, with k as the variable of integration, represents the part of the area for which $K_s < r$. Hawkins and Cundy [1987] pointed out that Equation (9.6) can be analytically integrated for the special case of an exponential distribution. For other distributions numerical methods are required. Figure 9.13 presents the solution of this expression in scaled form for a range of values of CV_K, with both rainfall rate r and K_e scaled on areal mean ξ_K. For CV_K of 0, K_e =

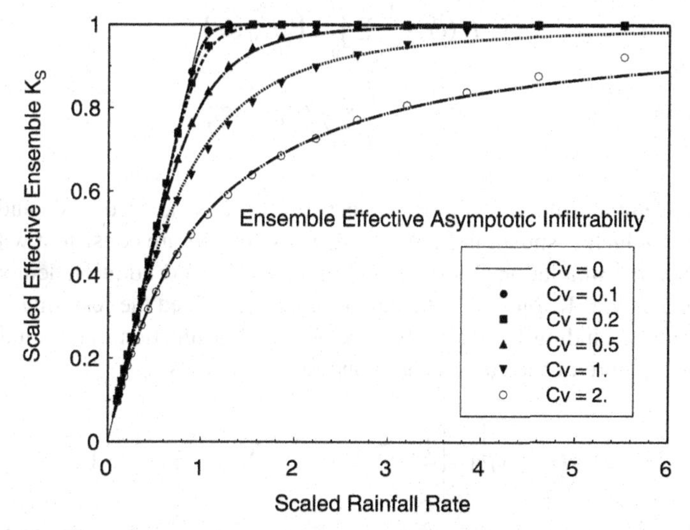

Figure 9.13. The existence of a continuous distribution of K_s across an area results in a relationship between the scaled rainfall rate and the effective areal mean asymptotic infiltration rate, K_e, which may fall significantly below the areal mean value of K_s.

ξ_K, the asymptotic value of infiltration rate is equal to r for $r < \xi_K$, and equal to ξ_K for all larger rain rates. With significant values of CV_K, the value of K_e is asymptotic to ξ_K as r increases, ever more gradually with increasing values of CV_K. Note that the bias in K_e is most marked at lower values of r_* and high values of CV_K. This is illustrated in comparing the effective K_e for Figures 9.12a and b.

Smith and Goodrich [2000] found that this relation could be very closely described by a family of empirical curves as follows:

$$\frac{K_e(r_*)}{\xi_K} = \left[1 + \left(\frac{1}{r_*}\right)^p\right]^{-1/p} \tag{9.7}$$

where r_* is r/ξ_K and

$$p = \frac{1.8}{CV_K^{0.85}} \tag{9.8}$$

Smith and Goodrich (2000) used Latin Hypercube ensemble simulation to investigate the dynamic relationship of ensemble f_c, referred to as f_e, as a function of ensemble I, referred to as I_e. In order to obtain a relation that is valid in time throughout a storm, the simulation was performed for an areal ensemble over time, and the relation of f_e and I_e obtained at equal times using the following:

$$I_e(t) = \frac{1}{n}\sum_{j=1}^{n}\int_0^t f\left(t, K_{sj}, G\right) \tag{9.9}$$

$$f_e(t) = \frac{1}{n}\sum_{j=1}^{n} f\left(t, K_{sj}, G\right) \tag{9.10}$$

Several ensemble simulations were made, using various values of CV_K and r_*, an example of which is shown in Figure 9.14. These results are consistent with previous ensemble simulation reports, as in Figure 9.12, and in addition demonstrate the dependency of K_e on r_*. Smith and Goodrich described the resulting infiltration relation, including Equations 9.7 and 9.8, with a function that included the transitional nature of the curve near "ponding" as follows:

$$f_{e*} = 1 + (r_{e*} - 1)\left\{1 + \left[\frac{(r_{e*} - 1)}{\gamma}\left(e^{\gamma I_{e*}} - 1\right)\right]^c\right\}^{-1/c} \quad ; \quad r_{e*} > 1 \tag{9.11}$$

Here r_{e*} is r scaled on K_e rather than ξ_K. This expression includes at its core Equation (6.25), including the parameter γ. In addition, the parameter c describes

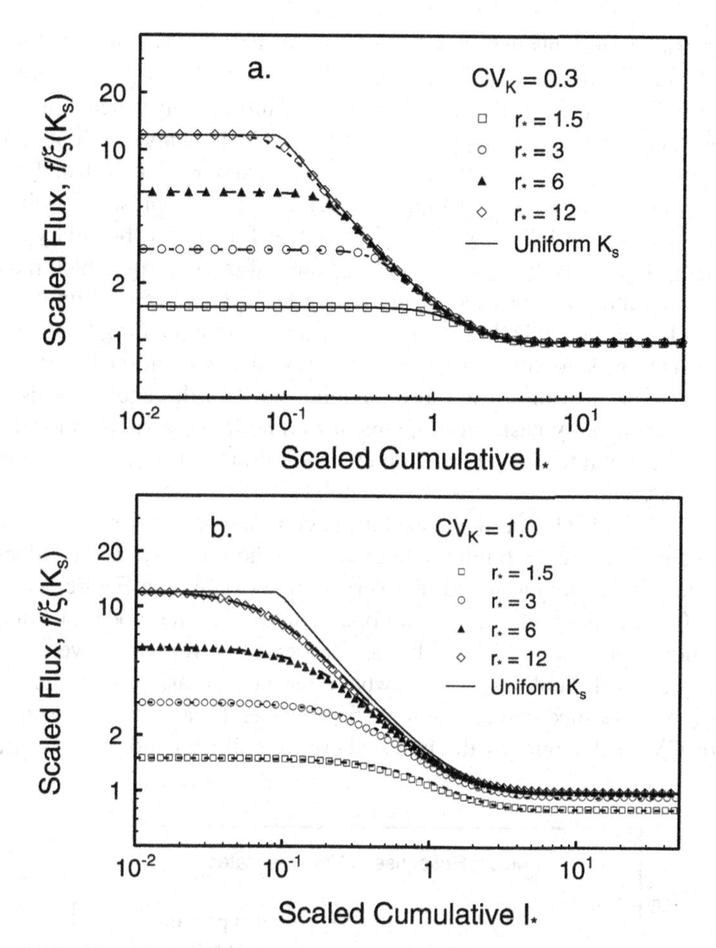

Figure 9.14 . Two values of CV_K and a range of rainfall rates illustrate the ensemble infiltration dynamics of an area with randomly varying K_s. Model Equation 9.11 is shown in each case by the lines which accompany the latin hypercube ensemble simulations sown with symbols.

the curvature of the transition, which is directly related to CV_K as illustrated in Figure 9.12. As c becomes increasingly large, the expression reduces to Equation (6.25). However, the scaling value used is K_e rather than $\xi(K_s)$. In this manner the partial area asymptotic effect is included as well, since K_e approaches $\xi(K_s)$ as CV_K goes to zero. To express the relation of c to r_* and CV_K, a single empirical expression was developed, based on a range of simulated ensembles:

$$c \cong 1 + \frac{0.8}{CV_K^{1.3}}\left[1 - \exp\left(-0.85(r_* - 1)\right)\right] \tag{9.12}$$

This expression represents the property that as scaled rainfall becomes sufficiently large, c in Equation (9.11) becomes dependent only on CV_K. Also, as CV_K becomes small, Equation (9.11) reduces to the core infiltrability function for $I > I_p$, and c approaches 1 for very large values of CV_K. Coefficients of variation of K measured in the field, however, have not been reported much larger than 2.

Another important property of Equation (9.11) is that it represents infiltration rate rather than infiltrability for all I_*. Recall that for rainfall boundary conditions, f_c from Equation (6.25) was equal to f only after ponding, which must be found by integrating r over time [using r for f in Equation (5.47)] until $f_c(I)$ is equaled. In Equation (9.11), however, there is no longer a ponding time, and the relation transitions smoothly from $f_c = r$ at $I = 0$ to the conventional infiltrability decay relation. Thus while it is more complicated than the equations listed in Table 6.1, it is actually easier to implement in a hydrologic runoff model. The only additional parameter required, implemented through the empirical relations developed, is CV_K.

While Equation (9.11) was developed using constant values of r, it is nevertheless applicable to unsteady rainfall rates, owing to the use of I_{e*} as the independent variable. This is demonstrated in a very simple example in Figure 9.15. The ensemble *LH* simulation is shown with open circles, and the model of Equation (9.11) is shown with the solid line. The case chosen is one for which, were CV_K=0 and $K_e = \xi_K$, runoff would begin somewhat after the increase of rainrate. This is illustrated by the dashed line in Figure 9.15. However, because K_e is less than ξ_K, runoff for CV_K = 1 begins at the point where rainfall steps up, as I_e is already

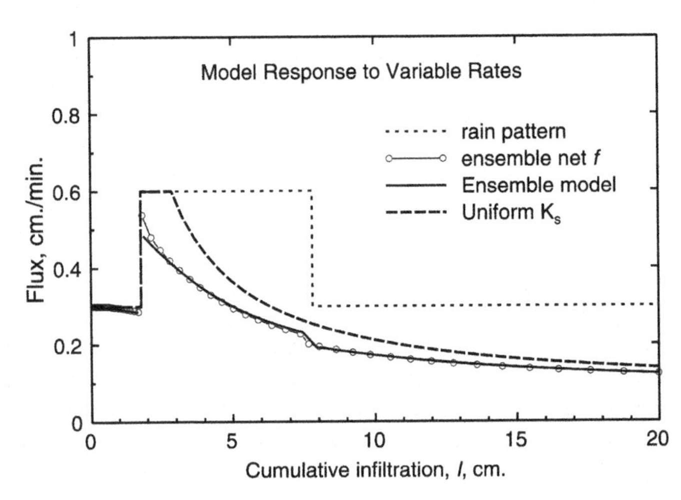

Figure 9.15. Demonstration of the effectiveness of the ensemble infiltration model in simulating the response to changing r values. The ensemble values from Latin Hypercube simulation are shown as open circles, and the solid line is the response simulated by Equation (9.11).

greater than the value of the increased $I_e(r_{e*})$ at this point. In this case f_e begins to drop below r almost immediately, but is negligible until the rainfall rate rises. The change in K_e when the rainfall rate subsequently decreases, through the relation of Equation (9.7), is modeled faithfully by application of Equation (9.11).

Smith and Goodrich [2000] found some field evidence that supported the use of this modified or large-scale infiltration relation. Unfortunately, there are almost no known careful rainfall and runoff measures on experimental catchments in which the random distribution of K_s is accurately known. The sampled data for distribution of K_s from the ARS R-5 catchment at Chickasha, OK, [Sharma, *et al.,* 1980] are not accompanied by reliable measures of local runoff pattern, due to instrumentation problems [Woolhiser, 1996]. But there is considerable evidence of the difference in areal effective K_e between lower rainfall rates and higher rainfall rates, as in Equation (9.7). This has been observed both in small experimental plots, as well as larger catchments [see Smith and Goodrich, 2000].

One-Dimensional Variability on a Catchment. The ensemble relationship of Equation (9.11) can be applied in a model that routes runoff interactively from a catchment. This approach neglects the possible effects of runon, or interaction of differing infiltrabilities along a flow path. Woolhiser and Goodrich [1988] used a parallel variation method to divide a one-dimensional catchment into n strips, oriented along flow paths, with a *LH* sample K_s assigned to each strip. This approach simulates parameter variation normal to flow, but does not treat variation along the flow path. These two methods might be expected to act quite similarly, insofar as runon is neglected. One can argue that the flow along a runoff path can have the effect of diminishing effective flowpath variability, but runon effects can have no such effect across adjacent flow paths. The strip method was indeed demonstrated to improve the hydrologic simulation of small catchments at Walnut Gulch Experimental Watershed [Woolhiser and Goodrich, 1988].

Smith *et al.* [1990] demonstrated some of the differences between the above types of ensemble approaches in simulating runoff response on heterogeneous catchments. When runoff was small compared with rainfall, the differences between methods was large. Conversely, it matters less which method is chosen when r_* is large and runoff begins quickly. This method will be compared with others below.

Two-dimensional Spatial Sampling The Smith *et al.* [1990] comparative study included the simulation of variations along and across flow paths by using a Monte Carlo method to create a random value of K_s at each numerical solution node along a flow path. The difficulty with a Monte Carlo Sampling method combined with a numerical solution for the surface flow dynamics is that there is a conflict between the number of nodes required for accurate simulation of the surface water flow equations, and the number of random samples needed to reduce sampling error to an acceptable level. Commonly, overland flow distances on catchments are on the order of 100 meters at the most. With this length, accurate kinematic wave simulation of runoff response can be made with as few as 12 to 15 node subdivisions. Thus the scale of Δx is on the order of 5m (or less). A random sample size of 15 is quite small to properly characterize a parameter distribution, and moreover, the scale of random variation in infiltration parameters such as K_s may be much smaller than this. Use of Latin Hypercube simulation can successfully simulate variations across paths, but accurate simulation of random variation along the flow path remains a challenge. For example, if variation along the flow path is simulated by random sampling, using only a few flow strips will cause bias due to the fact that there are a large number of possible arrangements of K_s along a flow path, and the arrangement [e.g. large K_s above a point with small K_s, or vice-versa] may be crucial for runoff dynamics.

A method for two-dimensional sampling that minimizes this bias is used here to demonstrate the effect of random variations in both dimensions. The spatial distribution of K_s is simulated using the *LH* method with n equal to the number of nodes required for kinematic wave numerical solution to Equations (9.4) and (9.5). Then a statistically large number of sample orderings are created having a random arrangement of these n values. Each sample arrangement is assigned to a flow strip and runoff simulated on the ensemble of strips. In this manner, the ordering of the *LH* samples of K_s is sufficiently randomized that the order along flow paths is not dominated by small sample statistical bias. For $n = 10$ *LH* samples, it was found that 50 Monte Carlo re-ordered strips was sufficient to reduce sampling order error bias to less than about 10%.

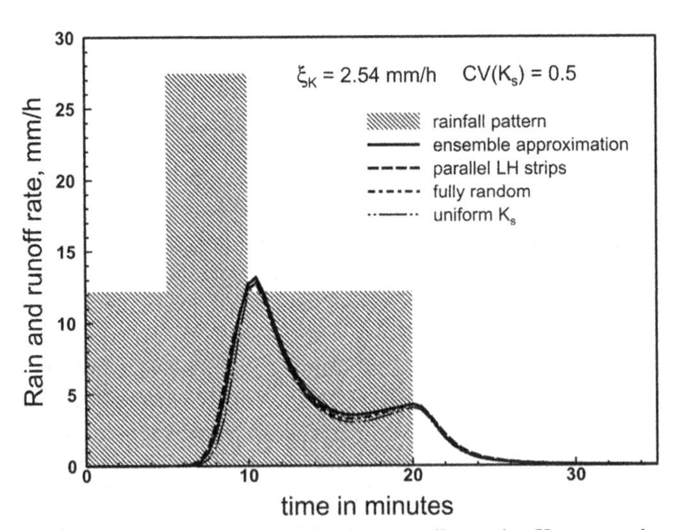

Figure 9.16. Three approaches to representing lognormally varying K_s on a catchment are compared here for a relatively large rainfall rate and a relatively moderate coefficient of variation of K_s. Under these conditions, the random variation of K_s does not make a significant difference.

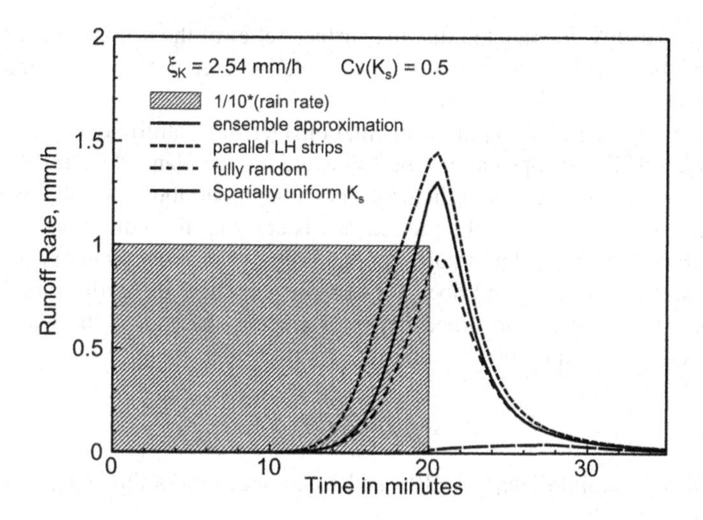

Figure 9.17. For the same random variability of K_s as in Figure 9.16, a smaller rainfall rate and relative runoff demonstrates more striking differences in the approaches for representing variations in K_s on a catchment, as illustrated here for a simple rainstorm pattern of 10 mm/h.

Figures 9.16 and 9.17 illustrate results for the comparison of three methods of treating infiltration parameter (K_s) variation discussed above. As can be seen in Figure 9.16, larger runoff cases can barely distinguish the 3 methods of approximation in simulating runoff response. However, in Figure 9.17, when r_{e*} is smaller or when runoff becomes a smaller proportion of rainfall, the sensitivity to the method of approximation for K_s distribution becomes significant. From this figure, one may conclude that any of the three methods of approximation is better than to assume a uniform K_s.

INFILTRATION AND "RUNON" EFFECTS

The runoff-runon phenomenon occurs whenever a more permeable soil area is located downslope from an area that generates surface runoff earlier, either because it has a low hydraulic conductivity or because shallow soils become saturated. This is a common occurrence in urban areas where runoff from roofs, sidewalks or streets may runon to grassed areas. In rural or natural settings it may occur when a soil disturbed by cultivation or compaction is upslope from pasture, hay or a vegetated buffer zone. It also occurs in mountainous regions or within arid or semi-arid climatic zones where low intensity winter rains can cause runoff from bare rock areas or can saturate shallow soils. This runoff can then be concentrated by infiltrating into deeper soils downslope. This redistribution of precipitation has important implications for groundwater recharge, and affects the location and growth rate of plant species.

It can be readily appreciated that the surface relief of the runon area is a critical factor, because it governs not only the relative area with ponded and infiltrating water but also the hydraulics of flow.

To illustrate the runoff-runon problem, consider the simplified case shown in Figure 9.18. This hillslope can be approximated by two plane elements. We consider the particular case where the upper plane is impervious and the lower area has a deep and porous soil. The rainfall rate is constant for a duration, D, and is smaller than the saturated hydraulic conductivity of the lower plane. If the lower plane is sufficiently long and does not become saturated, the entire runoff from the upper plane will be absorbed before it reaches the end of the plane. This runoff plus the rainfall will be called F_{max} :

$$F_{max} = PD(A_1/A_2 + 1) \tag{9.13}$$

where PD is the rainfall depth and A_1 and A_2 are the areas of the upper and lower planes respectively.

As soon as the rainfall begins, runoff flows from plane 1 to plane 2 and will proceed across it as an advancing front. For kinematic routing this front is most accurately treated as a kinematic shock (see Appendix 2). This is identical to the advancing front of surface irrigation during rainfall. The effect of surface relief on the velocity of the front and infiltration can be demonstrated by assuming a geometrical abstraction for the surface of plane 2 consisting of parallel, vee-shaped channels with spacing, W, and relief, RE as shown in Figure 9.19a. During the advance phase, infiltration proceeds at infiltrability over the area covered by water as shown by the cross-hatched area in Figure 9.19b and at the rainfall rate over the remaining area. For RE = 0 the entire portion of the plane behind the advancing front infiltrates at infiltrability. The rate of advance, as reflected by the time delay in runoff, the flow depth and the proportion of plane 2 covered with water all depend on the inflow rate, slope, hydraulic roughness

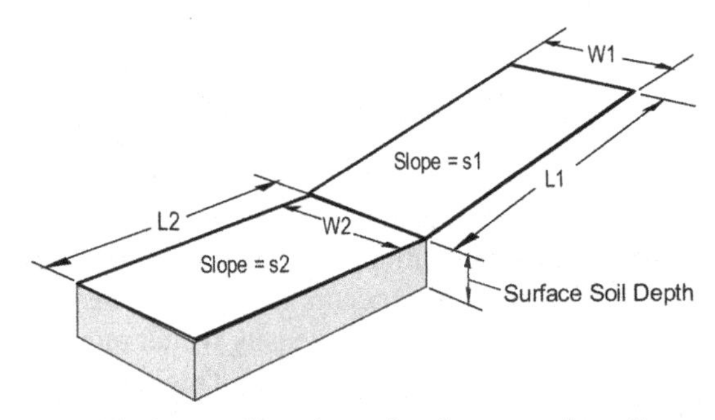

Figure 9.18. Hypothetical cascading planes for demonstration of runon hydrology. The upper plane is impervious, for the simplified example discussed in the text, and the rainfall rate does not exceed the infiltrability of the lower area.

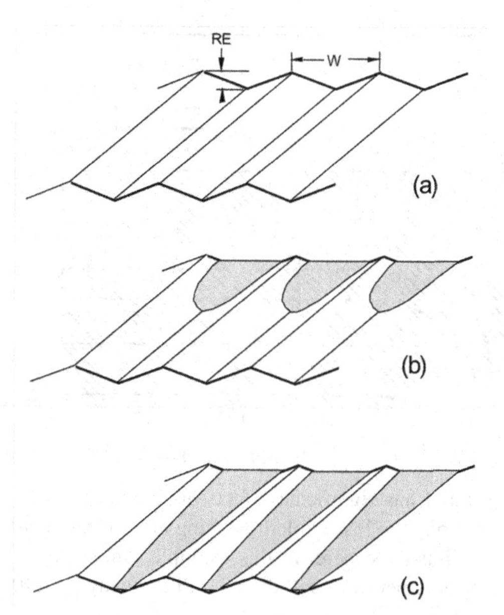

Figure 9.19. (a) the simplified microrelief geometry is defined as a vee-channel of width W and depth RE, (b) the runon flow advances like an irrigation advance, and with constant loss, approaches a steady non-uniform profile as in (c).

and RE. For a steady rainfall rate, the flow configuration on plane 2 would approach a steady state as shown in figure 9.19c, with a monotonic downslope decrease in the runoff rate due to infiltration losses.

An example of the effect of RE on the runoff hydrograph is shown in Figure 9.20. The total rainfall was 25.4 mm with a duration, D of four hours. For RE = 0 the advancing front of runoff never reaches the lower boundary of plane 2, so all of the runoff from plane 1 is infiltrated in addition to the direct rainfall on plane 2. As RE increases from 5 mm to 150 mm, the advancing front moves more rapidly and the runoff volume increases, reflecting the greater concentration of runoff. We define the *infiltration excess* on the runon plane as that depth of water that infiltrates from the runon, in excess of the direct rainfall. The infiltration excess in plane 2 varies from 25.4 mm for RE = 0 to 3 mm for RE = 150. Although infiltration excess is affected by many other variables, it appears that RE and the rainfall duration have a major influence. The effect of rainfall duration is shown in Figure 9.21 where an increase of the rainfall duration, while retaining the same volume, increases the infiltration excess to 25.4 mm for RE = 0, 5 and 10 mm. For RE = 150 mm the infiltration excess has increased from 3 mm to 11.25 mm.

Although the model described here is a crude approximation of natural surfaces, it leads to the conclusion that the runoff- runon phenomenon can be very important under some circumstances. It would be most significant in arid or

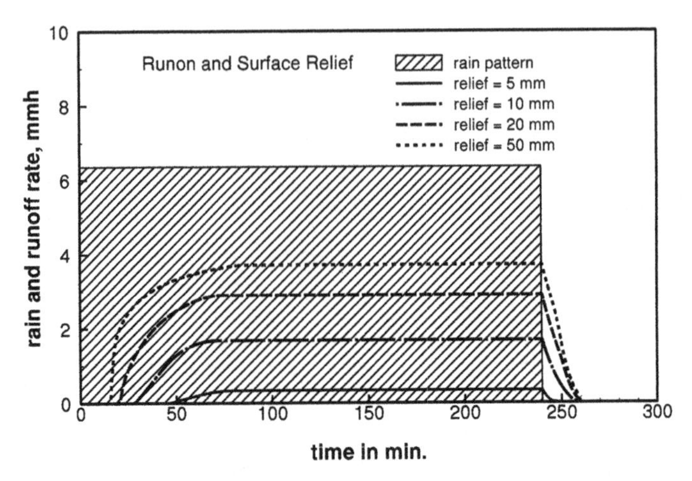

Figure 9.20. Surface relief has significant effects on infiltration for a case where runoff from an upper area is subject to loss while traversing an infiltrating area prior to entering a receiving stream, as illustrated here. In this example, runoff becomes nearly constant because the upper area is impervious and the rainfall is steady for 240 min.

semiarid climates where runoff is generated by Hortonian processes or by saturation of shallow soils on upslope areas. It would be relatively unimportant for undisturbed humid regions where most of the runoff is generated by saturation of areas adjacent to stream channels. Any watershed model, whether based on a

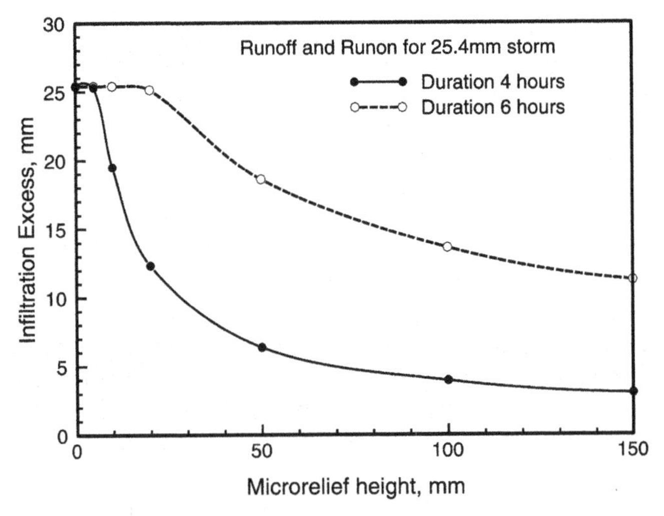

Figure 9.21. Effect of the duration of a given rain depth and the downslope microrelief on the infiltration excess of a storm when runon is an important factor.

cascade of planes, flow strips or pixels can be subject to serious error if this phenomenon is ignored. Additional research on the runoff-runon phenomenon should be encouraged.

SUMMARY

In this chapter we have touched on several aspects of the interaction of runoff and infiltration fluxes that may be important in hydrologic science and engineering. In many hydrologic analyses and hydrologic models, these interactions are ignored. This is especially significant in the adoption of hydrologic models to GIS simulation technology. While the use of GIS software allows the spatial representation of topography, soils, and other important hydrologic variables, the hydrologic dynamics incorporated is most often of a rather primitive nature. To some extent this is the result of the inability of many GIS software packages to allow for sophisticated and nonlinear dynamics in spatial interactions and spatial runoff analysis. But another significant limitation is in the detail of knowledge of spatial variations within a catchment of hydrologic interest. The availability of sophisticated GIS software and fast computers has not overcome limitations due to the difficulty and/or expense of gathering field information on variations in hydrologic soil properties affecting infiltration and runoff. Nor can remote sensing determine local values of many of the most important hydrologic soil parameters.

Appendices

I. DERIVATION OF THE INFILTRATION INTEGRAL FROM RICHARDS' EQUATION.

First, the Richards equation in θ - form:

$$\frac{\partial \theta}{\partial t} = \frac{\partial}{\partial z}\left(D \frac{\partial \theta}{\partial z} \right) - \frac{dK}{d\theta}\frac{\partial \theta}{\partial z} \quad \text{(A1.1)}$$

is transformed by use of the chain rule from partial differential equations:

$$\left(\frac{\partial q}{\partial t} \right)_z \left(\frac{\partial z}{\partial q} \right)_t = -\left(\frac{\partial z}{\partial t} \right)_q \quad \text{(A1.2)}$$

Multiplying both sides of equation (A1.1) by $\partial z/\partial \theta$ and applying the chain rule when appropriate:

$$-\frac{\partial z}{\partial t} = \frac{\partial}{\partial \theta}\left(D \frac{\partial \theta}{\partial z} \right) - \frac{\partial K}{\partial \theta} \quad \text{(A1.3)}$$

Then this expression is first integrated from θ_i to θ to obtain:

$$-\frac{\partial}{\partial t} \int_{\theta_i}^{\theta} z(\vartheta,t)\, d\vartheta = D \frac{\partial \theta}{\partial z} - (K - K_i) \quad \text{(A1.4)}$$

and rearranged:

$$\frac{\partial z}{\partial \theta} = \frac{D(\theta)}{K(\theta) - K_i - \frac{\partial}{\partial t}\int_{\theta_i}^{\theta} z(\vartheta,t)\, d\vartheta} \quad \text{(A1.5)}$$

Infiltration Theory for Hydrologic Applications
Water Resources Monograph 15

The differential-integral term in the denominator is simply the flux $v(\theta, t)$, as illustrated in Figure 5.1, and this expression can be again integrated now from θ to the surface value θ_o to obtain

$$z(\theta) = \int_{\theta}^{\theta_o} \frac{D(\theta)\,d\theta}{v(\theta) - (K - K_i)} \tag{A1.6}$$

where the signs in the denominator are changed along with the sign of z in the left side. Integration again with respect to theta over the entire wetted zone will change the left side from z to *I*. The right side is integrated by parts, and the flux concentration relation $F = v/v_o$ is used to represent $v(\theta)$ as a function of the surface flux, v_o:

$$I(t) = \int_{\theta_i}^{\theta_o} z\,d\theta = \int_{\theta_i}^{\theta_o} \frac{(\theta - \theta_i) D\,d\theta}{F v_o - (K - K_i)} \tag{A1.7}$$

Parlange [1980] takes a somewhat different approach to the above integration procedure [Also see Parlange *et al.*, 1982]. He integrates Equation (A1.4) with respect to z and obtains

$$\frac{\partial}{\partial t} \int_{\theta_i}^{\theta_s} \frac{-z^2}{2} d\theta = \int_{\theta_i}^{\theta_s} D\,d\theta - \int_{\theta_i}^{\theta_s} (K - K_i) \frac{\partial z}{\partial \theta} d\theta \tag{A1.8}$$

and then substitutes Equation (A1.5) for the $\partial z/\partial \theta$ term on the right, and replaces the first term on the right with $S^2/2/(\theta_s - \theta_i)$. He also assumes that the wetting front is close enough to a step function that the left side can be taken equal to $\frac{1}{2} I' f_c / (\theta_s - \theta_i)$. This form of the equation includes separate terms for the absorption and gravity effects.

II. ATTENUATION OF CHARACTERISTIC KINEMATIC SHOCK IN SOIL WATER FLOW

Assume kinematic wave flow in a soil with a power relationship between K and θ such that $K(\theta) = K_s(\Theta_e)^\varepsilon$, where Θ_e is reduced or normalized water content, Equation (2.5):

$$\Theta_e = \frac{\theta - \theta_r}{\theta_s - \theta_r}$$

The celerity of a θ 'wave' moving in response to gravity has a characteristic velocity u_c:

$$u_c = \varepsilon K_s \Theta_e^{\varepsilon - 1} \qquad \text{(A2.1)}$$

However, the translation velocity of a step increase in θ moves as a 'shock' as described by Equation (3.27). Figure 3.3 illustrates the graphical definitions of the shock velocity and the characteristic velocities for different values of θ.

For illustration, assume a simple square wave of maximum value θ_m is formed such that at some time t = 0 the input flux ceases and the surface condition reverts to the initial $\theta = \theta_i$, with the trailing 'step' values of θ from θ_i to θ_m. We may take initial water content equal to residual for simplicity. Any initial θ could be assumed with only inserting a constant into the calculations. Figure A1 illustrates the various key times in the attenuation of this simple "square" wave. The characteristic at θ_m, which is moving more rapidly than the shock front, will reach the advance shock wave at some time t_o and depth z_o. The time of intersection is dependent on the volume in the initial square wave, and affects the initial conditions for attenuation and the rate of attenuation, as will be shown below.

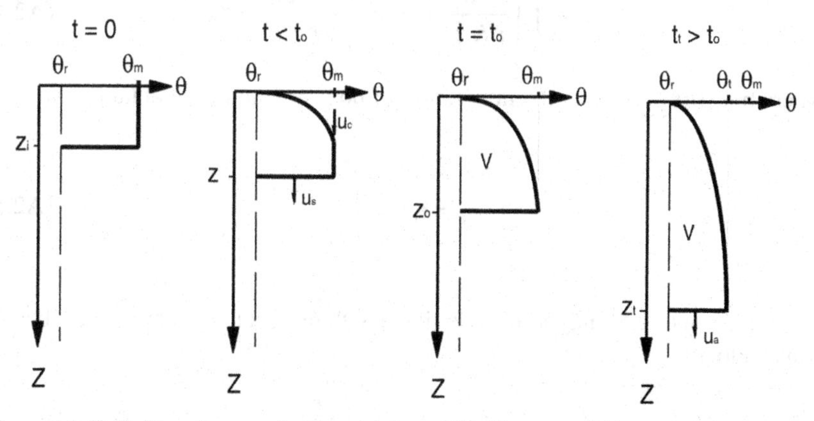

Figure A1. Definition diagram for the merging of "trailing wave" characteristics for kinematic soil water flow with the advance or kinematic "shock" wave. this occurs at time t_o, after which the pulse of soil water attenuates with time and depth.

The profile at times $t < t_0$, up to the depth of the shock, z_m, is described by the characteristic velocity and time:

$$z(\theta) = u_c(\theta)t = K_s \varepsilon \Theta_e^{\varepsilon-1} t \tag{A2.2}$$

which may be inverted to

$$\Theta_e(z) = \left(\frac{z}{\varepsilon K_s t} \right)^{\frac{1}{\varepsilon-1}} ; \quad \varepsilon > 1 \tag{A2.3}$$

Recall (Equation 3.27) that up to time t_o the advance wave is a shock which is moving at rate u_s:

$$u_s = \frac{K_s\left(\Theta_{em}^{\varepsilon} - \Theta_{ei}^{\varepsilon}\right)}{\theta_m - \theta_i}$$

Attenuation will begin from this point, as the advance wave can no longer move at the shock velocity, since the characteristics near the peak cannot travel beyond the shock. We may calculate the subsequent attenuation of the wave "peak", θ_t, by conservation of the volume of water in the wave. We equate the volumes, V, of the wave for all $t \geq t_o$. Taking θ_i to be near to θ_r, the characteristic at the upper base of the wave will not move, so we may integrate Equation (A2.3) over the depth 0 to z:

$$V = \int_0^{z_o} \left(\frac{z}{\varepsilon K_s t_o} \right)^b dz = \int_0^{z_t} \left(\frac{z}{\varepsilon K_s t} \right)^b dz \tag{A2.4}$$

where b is 1/(ε -1). From the integration of both functions, we obtain

$$z_t = z_o \left(\frac{t}{t_o} \right)^{\frac{1}{\varepsilon}} \tag{A2.5}$$

Equation (A2.5) may be differentiated with respect to t to obtain the advance wave velocity:

$$u_a = \frac{dz}{dt} = \frac{z_o}{\varepsilon t_o} \left(\frac{t_o}{t} \right)^{\frac{\varepsilon-1}{\varepsilon}} ; \quad t > t_o \tag{A2.6}$$

The new wave advance velocity is then different from the associated shock velocity by the ratio of $(\Theta_e)^{\varepsilon-1}$ to $(\Theta_e)^{\varepsilon}$. Also, noting that at any time $z > z_o$ the velocity of the water content at the attenuating peak, θ_t, is still described by the characteristic originating from the initial time = 0, one may combine the relation (A2.3) for characteristic depth and θ with Equation (A2.5) to obtain the attenuation as a function of time:

$$\Theta_{et} = \Theta_{eo}\left(\frac{t_o}{t}\right)^{\frac{1}{\varepsilon}} \tag{A2.7}$$

One may also differentiate Equation (A2.7) to obtain the attenuation rate of the peak, substituting the relation of θ to Θ_e [Equation (2.5)]:

$$\frac{d\theta_t}{dt} = \frac{-(\theta_o - \theta_r)}{\varepsilon t_o}\left(\frac{t_o}{t}\right)^{\frac{1+\varepsilon}{\varepsilon}} \tag{A2.8}$$

Finally, Equation (A2.8) may be combined with Equation (A2.6) to obtain the attenuation of the peak θ_t with depth:

$$\frac{d\theta_t}{dz} = \frac{\frac{d\theta_t}{dt}}{\frac{dz}{dt}} = -\frac{(\theta_o - \theta_r)}{z_0}\left(\frac{t_o}{t}\right)^{\frac{2}{e}} \tag{A2.9}$$

Similar results, with complicating terms, may be obtained for the case where the water content, and thus the flow, in advance of the wave and at its "tail" are nonzero.

III. FINITE DIFFERENCE SOLUTION METHOD FOR RICHARDS' EQUATION

In a finite-difference solution of a differential or partial differential equation, both space and time coordinates are divided into finite increments. Individual points, or *nodes*, are numbered i = 1,2,...n, separated by a uniform or variable incremental distance, referred to as, for example, Δx, Δz, or Δt. The differential dx becomes the difference Δx, which can also be referred to as $x_i - x_{i-1}$. Depending on the formulation, the nodes may conceptually be at the centers or at the boundaries of the finite increments.

Forming the finite difference equations. In the following expressions, the superscript index *j* will be used for time increments, and subscript *i* will be used for spatial increments. We use the convention that z is measured positive downward from the surface, and the nodes are numbered in tis direction also. With this finite difference transformation, Equation (3.15) may be written in the simplest terms as:

$$\frac{\left[\theta^{j}-\theta^{j-1}\right]}{\Delta t}= \frac{-2}{\left(z_{i+1}-z_{i-1}\right)}\left\{\begin{array}{l}\omega\left[\overline{K}_{i,i+1}\left(\frac{\left(\psi_{i+1}-\psi_{i}\right)}{\Delta z_{i,i+1}}-p\right)-\overline{K}_{i,i-1}\left(\frac{\psi_{i}-\psi_{i-1}}{\Delta z_{i,i-1}}-p\right)\right]^{j}+ \\ (1-\omega)\left[\overline{K}_{i,i+1}\left(\frac{\left(\psi_{i+1}-\psi_{i}\right)}{\Delta z_{i,i+1}}-p\right)-\overline{K}_{i,i-1}\left(\frac{\psi_{i}-\psi_{i-1}}{\Delta z_{i,i-1}}-p\right)\right]^{j-1}\end{array}\right\} \tag{A3.1}$$

The symbol p is the gravitational term here equal to 1 for vertical infiltration, and this equation can also represent horizontal flow, Equation (3.2), for p = zero . The right side is complicated by the time-weighted treatment of spatial differences, with weighting facter ω. When this weighting is 1.0, this becomes a fully implicit form that requires iterative solution, or if ω is 0, the spatial differences at the previous time level j -1 are used, allowing an explicit solution, dependent only on values known from the previous time. There are many formulations for weighting, and for determination of mean coefficient values $\overline{K}$, all of which reduce to the differential equation as the increment size becomes infinitesimally small. The art in numerical solution is to find an efficient method to establish these coefficient values and the time weighting for the spatial difference of the right side of this form between level j and j -1. The equal weighting form, $\omega = ½$ is called the *Crank-Nicholsen* formulation. Another complication in the solution of the difference equation is that the left side contains an unknown value of θ_i^j, while the right side is in terms of unknown ψ values (at i-1, i, and i+1). Some solutions cast the equation in terms of only ψ by using the retention relation with the time differential on the left side including the 'specific moisture capacity,' C_r:

$$\frac{\partial \theta}{\partial t} = \frac{d\theta}{d\psi}\frac{\partial \psi}{\partial t} = C_r(\psi)\frac{\partial \psi}{\partial t} \tag{A3.2}$$

However, experience has shown that selection of C_r is critical to conservation of mass. This was discussed in some detail by Celia et al.(1990). Their solution, often used since, consists of expanding $\theta(\psi)$ in a Taylor series about Θ^j. Such complications are felt to be unnecessary for mass conservation, as long as it is realized that $\theta(\psi)$ may be retained in the implicit numerical formulation. This in effect defines C_r as the chord slope of the retention relation (rather than the tangent at some unknown point) between the values of ψ at j and j-1 (see figure A2). Thus the change in θ is kept perfectly consistent with the change in ψ in the left side of the equation.

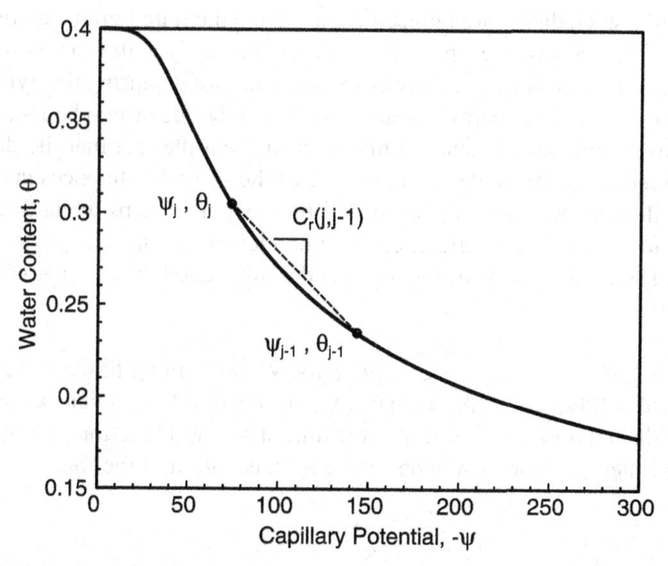

Figure A2. Definition of the effective soil water capacity value, C, used in the numerical solution for a location in the soil where the state of the soil, ψ and θ , is moving from point j-1 to j during a time step. Use of a mean or weighted tangent slope rather than the chord slope generally results in significant errors of mass balance.

The selection of $\overline{K}$ representing the effective hydraulic conductivity for flow between node points i and i-1, is important especially for wetting front conditions where there is a steep gradient of ψ and thus rapidly varying values of K. This has been the subject of several published papers [Haverkamp and Vauclin, 1979; Zaidel and Russo, 1992]. The numerical solution used here is consistent with the findings of Zaidel and Russo [1992] in the use of a weighting based on the assumption of a linear change of ψ between nodes:

$$\overline{K}_{i,i-1} = \frac{1}{\psi_i - \psi_{i-1}} \int_{\psi_{i-1}}^{\psi_i} K(h)dh \tag{A3.3}$$

Here h is the variable of integration. This method captures the strong nonlinearity of the K(ψ) relation much better than either arithmetic or geometric means, which others have used [Haverkamp and Vauclin, 1979].

Increment size may vary within the solution space, and for solution of the adsorption or infiltration case, it is an advantage to have intervals that are smallest in the region of solution where the gradients are steepest, i.e., near the entrance boundary. For infiltration from a surface boundary, the increments should be smallest near the surface and may increase slowly at larger values of z (or x) as the ψ gradients are reduced compared to the initial gradient conditions. For the same reasons, the time steps may be increased with time as flow rates become smaller. In any case, there is no possibility of capturing the dynamics of infiltration during time periods near zero when only one or two boundary increments have been wetted. This should be clear from the fact that the difference equation requires three nodes for definition of the second differences in the equation. While mass balance can be assured for early time steps while these nodes are wetting, a numerical difference or finite element solution approaches accurate flux simulations only after several nodes are wetted, however small may be Δz and Δt.

Solving the set of difference equations. Without going through the detailed steps involved, Equation (A3.1) may be rearranged to form a matrix equation in terms of the unknown values of ψ at the time step j, as a function of those known from time step j-1. For each node i there is an equation of the form

$$a_{i-1}\psi_{i-1}^{j} + a_i\psi_i^{j} + a_{i+1}\psi_{i+1}^{j} = B_i^{(j-1)} \tag{A3.4}$$

The coefficients a are functions of the variables ψ and form a coefficient matrix that is composed of a major diagonal and two adjacent minor diagonals. The vector B comprises values of terms known from the previous time step. Solution of this set of linear equations in ψ is often undertaken by repeated successive trials, known as Picard iteration. A more efficient method is to use the Newton-Raphson method for systems of equations. Briefly, Equations (A3.1) are formed into a set of objective functions whose values over the solution space, nodes i = 1,N are to be minimized:

$$F(\psi_i^j) = 0 \qquad \text{(A3.5)}$$

The Newton-Raphson solution for this set of equations (i=1,N) consists of solving for the correction term vector, $\delta(\psi)$, in the set of equations

$$F'(\psi)\delta(\psi) = F(\psi) \qquad \text{(A3.6)}$$

where F' is $dF/d\psi_i$. These terms form a tri-diagonal matrix since F_i is a function of values from nodes i-1, i, and i+1. The procedure is recursive, with ψ_i corrected by $\delta(\psi_i)$ at each iteration until the largest value of the objective function vector F, or the correction vector δ, is less than preset criterea.

Boundary conditions are used to write equations for the first and last row of the N equations involved. At the soil surface, the finite difference equivalent of Equation (3.17), during rainfall r prior to surface ponding, may be written

$$\omega\left\{\overline{K}_{1,2}\left[\frac{\psi_1-\psi_2}{\Delta z_{1,2}}+p\right]\right\}^j+(1-\omega)\left\{\overline{K}_{1,2}\left[\frac{\psi_1-\psi_2}{\Delta z_{1,2}}+p\right]\right\}^{j-1}=r \qquad \text{(A3.7)}$$

Again, coefficients are functions of the unknown adjacent values of ψ. After the time when the solution finds a surface value of ψ equal to or greater than 0, the upper boundary equation becomes simply $\psi_1 = 0$.

Lower boundary conditions may be formulated in more than one way. In general, solutions shown in this work have used cases where the lower boundary is not reached by water entering at the surface, so the lower boundary may be simply a fixed head: $\psi_N(t>0) = \psi(t=0)$. Alternatively, a lower boundary may be formulated by specifying gravity drainage, or no flow, or a water table (which is also a fixed head condition). In each case, a finite difference equation in one or two ψ terms (nodes N and N-1) such as Equation (A3.7) may be written.

Selection of Time Steps. The solution efficiency is sensitive to selection of time steps. Steps too small simply extend the work of solution, and those too large make convergence more difficult, at least. The method employed here uses a criterea based on the flux conditions at the previous time step, and calculates the time step size based on the flow rate into or out of each finite node volume. The fluxes into and out of all nodes are calculated, and maximum relative potential storage change $\Delta V_i/V_i$ is found. If this is positive, it is compared with local available storage $(\theta_s - \theta)$, and when negative, it is compared with $(\theta - \theta_r)$; the time step is chosen to match these preset limits.

Validation Equations (4.43-4.46) may be used, as indicate earlier, to validate numerical models such as described here, that can then apply to many cases that do not obey the exact limitations of the analytic solution. The numerical model used here is compared with this analytic solution in Figure A3, for a relatively nonlinear soil with $C_n = 1.01$, $\lambda_s(=G) = 20.$, and normalized rainfall rate of 2.5. The solution shown here is at 1 hour after start of rainfall, after about 4000 iterations of the numerical model. The spatial increments vary from 0.5mm near the surface to 5mm at depths of 200 mm. or greater. Close examination reveals possibly some small numerical diffusion in the most advanced part of the wetting wave where water contents rise at very low conductivities, but overall the performance of the numerical model is excellent.

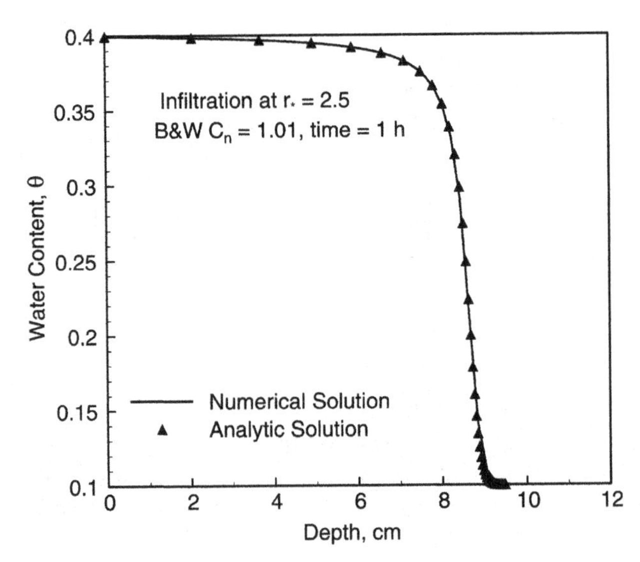

Figure A3. The numerical method of solving Richards' equation described in the Appendix is here verified using the Broadbridge-White solution described in Chapter 4. If one looks very closely, there is evidence of possible numerical dispersion, not surprisingly, just at the most forward part of the advancing wave.

IV. COMPUTATION OF THE SOLUTION TO EQUATIONS (4.43)-(4.46)

The erc(-) function

The function *erc*(-) that was introduced for the integrable soil of Broadbridge and White [1988] arises many times in the solution unsaturated flow models, and, was defined in Chapter 4:

$$erc(x) = exp(x^2)\, erfc(x) \qquad 4.16$$

In this sense, it is a more practical function than the complementary error function. The function *erc*(x) is graphed in Figure A4. It has a relatively simple asymptotic expansion [Abramowitz and Stegun , 1964] since

$$\sqrt{\pi} x e^{x} 2erf(x) \approx 1 + \sum_{m=1}^{\infty} (-1)^m \frac{1 \cdot 3 \cdots (2m-1)}{(2x^2)^m} \qquad (A4.1)$$

However, the numerical evaluation of *erc*(x) is the most common source of error in evaluating the analytic solutions. For values of x higher than 3 or so, we should not directly carry out the multiplication operation in (4.16), as this involves the partial cancellation of a very large number *exp* (x^2) by a very small number *erfc* (x), requiring many digits and leading to significant numerical errors. It is safer to directly approximate *erc*(x) as a rational function such as [Hastings, 1955; Gautschi, 1965]:

$$\begin{aligned} erc(x) &= a_1 t + a_2 t^2 + a_3 t^3 + a_4 t^4 + a_5 t^5, \quad t = 1/(1+px); \quad x \geq 0 \\ erc(x) &= 2exp(x^2) - erc(x)\,; \qquad x < 0 \end{aligned} \qquad (A4.2)$$

where p = 0.3275911
a1 = 0.254829592
a2 = -0.284496736
a3 = 1.421413741
a4 = -1.453152027
a5 = 1.061405429

A MATLAB Program for Computations

The following code was developed for MATLAB5 to solve the set of equations (4.43-4.46) constituting the analytic solution of Richards' equation for the Broadbridge-White soil characteristics. The code is relatively transparent and can be readily translated into such languages as C, FORTRAN, Basic, or others.

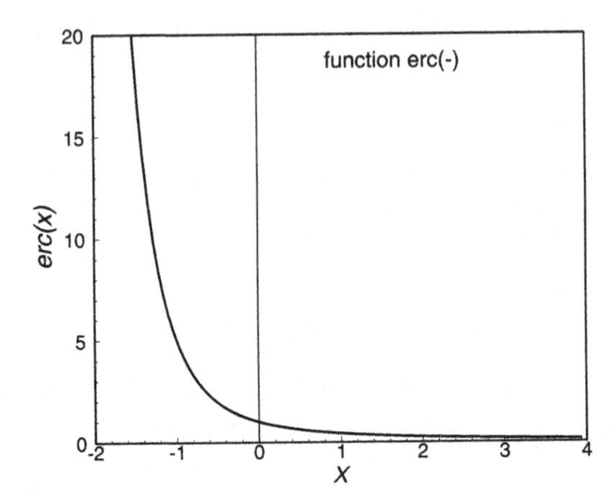

Figure A4. The mathematical function *erc*( ) is here graphed for the more interesting central part of its argument.

MATLAB5 program profile.m

```
% specify soil and rainfall data:
thn=.2376;                                              θn
ths=.495;                                               θs
thi=.3;                                                 θi
Thi=(thi-thn)/(ths-thn);                                Θi
% R = ratio of rainfall rate to Ks
R=1;                                                    R/Ks = 1
Cn=1.17;                                                Cn
% specify a dimensionless time
T=2;                                                    t*
rho=.25*R/(Cn*(Cn-1));                                  ρ
tau=4*Cn*(Cn-1)*T                                       τ
zemax=(Cn-1+1.5*(Cn-Thi))*R*T/(1-Thi);                  limit of ζ
zet=linspace(0,zemax,101);
% then find kappa:
ka=2*rho-Thi/(Cn-Thi);                                  κ

for j=1:101
ze=zet(j)                                               ζ
u1=exp(ka*ze+ka^2*tau/4;
a3=ka*sqrt(tau)/2;
a1=ze/sqrt(tau);
a2=sqrt(rho*(rho-1)*tau);
e1=exp(-1*a1^2);
u2=e1*erc(a1+a2)
u3=e1*erc(a1-a2);
u4=e1*erc(a1+a3);
u5=e1*erc(a1-a3);
u=u1+.5*(u2+u3-u4-u5);                                  u
uze=ka*u1+sqrt(rho+1))*(u2-u3)+(ka/2)*(u5-u4)           ∂u/∂ζ
B1=1/(1+2*rho-uze/u);
Thet(j)=Cn*(1-B1);                                      Θ
Z(j)=(rho*(rho-1)*tau+ze*(2*rho+1)-log(u));             z*
end
% plot the profile
plot(Z,Thet)
xlabel('Z')
ylabel('Theta')
%this creates a postscript file profile.ps:
print profile
%
% this is the erc function erc.m
function y=erc(x)
z=abs(x);
p=.3275911;
a1=.254829592;
a2=-.28449676;
a3=1.421413741;
a4=-1.453152027;
a5=1.061405429;
t=1/(1+p*z);
y=a1*t+a2*t^2+a3*t^3+a4*t^4+a5*t^5;
ifx<0
y=2*exp(z^2)-y;
end
```

List of Symbols and Abbreviations

Symbol	Meaning	Units
A_f	Relative surface area covered by water during flow recession-	
A_n	coefficient of *n*th term in Philip series solution	
C	locally defined coefficients	
C_n	nonlinearity parameter in Broadbridge-White soil	
$C_r(\psi)$	slope of the retention function: $d\theta/d\psi$	L^{-1}
c	(chap. 9) parameter of ensemble infiltration function reflecting CV_K or, (chap. 2) the curvature parameter in the T-BC soil characteristic function	
CV_K	coefficient of variation of K_s	
D	soil water diffusivity, $Kdh/d\theta$	L^2T^{-1}
e_r	rainfall excess, the value $f - r$ at the soil surface	LT^{-1}
f	infiltration rate, or value of v at the soil surface: $f \equiv v_o$	
f_c	infiltrability, the maximum rate that the soil can adsorb or infiltrate water at ψ_o=0 at current soil conditions	LT^{-1}
F	flux-concentration, relative variation of soil v throughout the wetting profile	
G	capillary length scale, or capillary drive parameter	L
h	depth of water on the surface, or used as variable of integration for ψ	L
H	total soil potential, capillary head plus gravitational potential	L
i	as subscript: to indicate initial conditions	
I	cumulative infiltration depth	L
I'	$I - K_i t$	L
I_p	value of I at ponding	L
IDA	Infiltrability Depth Approximation - see Chapter 5	
k_r	relative hydraulic conductivity, K/K_s	-
K	soil hydraulic conductivity	LT^{-1}
K_e	Areal effective value of K_s for use in ensemble infiltration function	
K_s	hydraulic conductivity at natural saturation	LT^{-1}
K_*	scaled hydraulic conductivity $(K - K_n)/(K_s - K_n)$ (Ch. IV)	
m	integer constant, or hydraulic exponent for surface water flow	

n	(Chapter 4) subscript denoting minimum (immobile) θ or scaling based on immobile θ	
o	subscript denoting soil surface	
q	unit discharge of surface water [discharge per unit width] or flux into infiltrometer (Chapter 8)	L^2T^{-1} L^3T^{-1}
q_∞	final discharge flux from surface pond (Chapter 8)	L^3T^{-1}
r	rainfall rate	L/T
r_*	normalized rainfall rate, scaled by K_s	
r_{e*}	rainfall rate normalized by K_e	
R	radius of permeameter or infiltrometer	L
S	sorptivity (Equation 3.12)	$LT^{-1/2}$
t	time	T
t_p	time of ponding; for flux boundary, the time when supplied flux first exceeds infiltrability	T
t_c	scaling or normalizing time (see Equation 6.2)	T
t_s	scaling time (Philip's t_{grav}) used in Chapter 4; $= ½\, t_c$	T
t_x	scaled time	
TB-C	Transitional Brooks-Corey soil type with characteristic functions described by Equations (2.20) and (2.21), comparable to the van Genuchten function, but retaining Brooks-Corey parameters	
u	soil water velocity = $v/(\theta - \theta_r)$	LT^{-1}
u_c	soil water celerity or characteristic velocity	
u.b.c.	upper boundary condition	
v	local soil flux. v_o is surface infiltration rate, equal to f_c or r, whichever is smaller.	LT^{-1}
x	distance, generally horizontal distance as opposed to vertical	L
z	depth, vertical distance measure, by convention here positive downward from the surface	L
Z	depth to bottom of wetted surface zone for delta-function case	L
α	parameter in the van Genuchten soil water retention formula	L^{-1}
α_G	parameter in the Gardner expression for hydraulic conductivity	L^{-1}
β	a) shape factor for describing wetting $\theta(z)$ curve (Chapter 7) b) weighting factor or constant in various equations	
γ	weighting factor used in the 3-parameter infiltration equation	
$\delta(\Theta_i)$	$\equiv F/\Theta_i$; marginal flux-concentration function-	
ε	exponent parameter on Θ_e in relation of k_r	

η	exponent parameter in relation of K to ψ; log slope of $K(\psi)$	
Θ_e	water content scaled between residual and maximum water content values; $= (\theta - \theta_r)/(\theta_s - \theta_r)$	
Θ_i	water content scaled on initial profile value $= (\theta - \theta_i)/(\theta_o - \theta_i)$	
Θ_n	water content scaled on minimum θ_n in analytic solution, Chapter 4	
θ	soil water content by volume	
θ_n	water content at minimum $K = K_n$ (Chapter 4)	
θ_o	surface water content	
θ_r	residual θ; water not removable by capillary forces	
θ_s	water content at $\psi = 0$.	
λ	log-slope of the soil water retention curve, often called the *pore-size distribution coefficient*	
λ_s	capillary length scale, essentially equal to G	L
$\xi(y)$	expected value of random variable y	
ξ_K	expected value of K_s	L/T
ρ	mass density	ML^{-3}
φ	Bolzmann transform variable, = distance scaled on inverse square-root of time	$LT^{-1/2}$
$\phi(h)$	flux potential, defined (Equation 2.7) as $\int_{-\infty}^{h} K(h)dh$	L^2T^{-1}
$\phi(a,b)$	differential flux potential between points a and b: or $\phi_{a,b} = \phi(a) - \phi(b)$	LT^{-1}
ψ	soil water capillary pressure head	L
ψ_a	presssure head shift parameter in the T-BC retention relation	L
ψ_B	pressure head intercept parameter, or 'air entry" parameter in the T-BC retention relation	L
ω	weighting parameter	

References

Ahuja, L.R., and D. Swartzendruber, An improved form of soil-water diffusivity function. *Soil Science Society of America J., 36*(1), 9-14, 1972.

Barry, D.A., and G.C. Sander, Exact solutions for water infiltration with an arbitrary surface flux or nonlinear solute adsorption, *Water Resources Research,* 27, 2667-2680, 1991.

Bear, J., *Dynamics of Fluids in Porous Media,* American Elsevier Publishing Co., New York, 1972.

Benton, E.R. and G. W. Platzman, A table of solutions of the one-dimensional Burgers equation, *Quart. Appl. Math.*, 30, 195-212, 1972.

Birkhoff, G. *Hydrodynamics: A study in logic, fact, and similitude,* Princeton University Press, Princeton, NJ, 1950.

Bluman, G.W., and S. Kumei, On the remarkable nonlinear diffusion equation $(\partial/\partial)[a(u+b)^{-2}(\partial u/\partial x)]-(\partial u/\partial t) = 0$, *J. Math. Phys., 21,* 1010-1023, 1980.

Bluman, G.W., and S. Kumei, *Symmetries and Differential Equations,* Springer, New Yourk, 1989.

Bouwer, H., Unsaturated flow in ground-water hydraulics, *J. Hydraulics Div., American Soc. of Civil Engr., 90* (HY5), 212-144, 1964.

Braester, C., Moisture variation at the soil surface and advance of wetting front during infiltration at constant flux, *Water Resour. Res., 9*, 687-694, 1973.

Bridge, B.J. and Ross, P.J. A portable microcomputer-controlled drip infiltrometer. II. Field measurement of sorptivity, hydraulic conductivity and time to ponding. *Aust. J. Soil Res. 23*, 393-404.

Broadbridge, P., The forced Burgers equation, plant roots, and Schrödinger's eigenfunctions, *J. Eng. Math., 36,* 25-39, 1999.

Broadbridge, P, and C. Rogers, On a nonlinear reaction-diffusion boundary-values problem: Application of a Lie-Bäklund symmetry, *J. Austral. Math. Soc. (b), 34*, 318-332, 1993.

Broadbridge, P., and J.M. Stewart, Time to Ponding During Non-Constant Irrigation, in *Engineering Mathematics: Research, Education and Industry Linkage* (W.Y.D. Yuen, P.Broadbridge, and J.M. Steiner, editors), pp. 469- 477, Institution of Engineers, Australia, 1996.

Broadbridge, P., and I. White, Time to ponding: Comparison of analytic, quasi-analytic, and approximate predictions, *Water Resour. Res., 23*, 2302-2310, 1987.

Broadbridge, P., and I.White, Constant rate rainfall infiltration: A versatile nonlinear model. 1. Analytic solution, *Water Resources Res., 24,* 145-154, 1988.

Brooks, R.H., and A.T. Corey, Hydraulic properties of porous media, *Hydrology Paper No. 3*, Civil Engineering Dept., Colorado State Univ., Fort Collins, CO, 1964.

Brutskern, R.L., and H.J. Morel-Seytoux, Analytical treatment of two-phase infiltration, *J Hydraul. Div, ASCE, 96*(HY12), 2535-2548, 1970.

Buckingham, E., Studies on the movement of soil moisture, *U.S. Dept. of Agriculture Bureau of Soils Bull. 38*, 1907.

Burdine, N.T., Relative permeability calculations from pore-size distribution data, *Petroleum Transactions, American Institute of Mining and Metallurgical Engineering, 198*, 71-77, 1953.

Burgers, J.M., A mathematical model illustrating the theory of turbulence, *Adv. Appl. Mech., 1*, 171-199, 1948.

Carslaw, H.S., and J.C. Jaeger, *Conduction of Heat in Solids,* 2nd edition, Oxford Univ. Press, London 1959.

Celia, M.A., E.T. Bouloutas, and R.L. Zarba, A general mass-conservative numerical solution for the unsaturated flow equation, *Water Resources Res., 26*(7), 1483-1496, 1990.

Charbeneau, R.J., Kinematic models for soil moisture and solute transport, *Water Resources Res., 20*(6), 699-706, 1984.

Chu, B.T., Parlange, J-Y., and Aylor, D.E. Edge effects in linear diffusion. *Acta. Mech.*, 21, 13-27, 1975.

Clothier, B.E., J.H. Knight, and I. White, Burgers' equation: Application to field constant-flux infiltration, *Soil Science, 132*, 252-261, 1981a.

Clothier, B.E., White, I., and Hamilton, G.J. Constant-rate rainfall infiltration: Field experiments. *Soil Sci. Soc. Amer. J.* 45, 245-249, 1981b.

Clothier, B.E., Green, S.R. and Katou, H. Multidimensional Infiltration: Points, furrows, basins, wells, and disks. *Soil Sci. Soc. Amer. J.*, 59, 286-292,1995.

Clothier, B.E., and Smettem, K.R.J., Combining laboratory and field measurements to define the hydraulic properties of soil. *Soil Sci. Soc. Amer. J.* 54, 299-304, 1990.

Cole, J.D., On a quasilinear parabolic equation occurring in aerodynamics, *Quart. Appl. Math., 9*, 225-236, 1951.

Cook, F.J., and Broeren, A. Six methods for determining sorptivity and hydraulic conductivity with disc permeameters. *Soil Sci.*, 157, 2-11, 1994.

Corey, A.T. *Mechanics of Immiscible Fluids in Porous Media,* Water Resources Publications, Highlands Ranch, CO, 1994.

Corradini, C., F. Melone, and R.E. Smith, Modeling local infiltration for a two-layered soil under complex rainfall patterns, *J. of Hydrology, 237*(1-2), 58-73, 2000.

Corradini, C., R. Morbidelli, and F. Melone, On the interaction between infiltration and Hortonian runoff, *J. of Hydrology, 204*,52-67, 1998.

Crawford, N.H., and R.K. Linsley, Digital simulation on hydrology, Stanford Watershed Model IV, Stanford University Technical Report No. 39, Palo Alto, CA, 1966.

Eagleson, P.S., *Dynamic Hydrology,* McGraw-Hill, New York, 1970.

Eagleson, P.S., Climate, soil, and vegetation 5. A derived distribution of storm surface runoff, *Water Resources Res., 14*(5), 741-748, 1978.

Edwards, M.P., and P. Broadbridge, Exact transient solutions to nonlinear diffusion-convection equations in higher dimensions, *J. Phys. A. Math. Gen., 27*, 5455-5465, 1994.

Elrick, D.E., and D.H. Bowman, Note on an improved apparatus for soil moisture flow measurements, *Soil Society of America Proceedings, 28*, 450-453, 1964.

Fokas, A.S., and Y.C. Yortsos, On the exactly solvable equation $S_t=[(\beta S+\gamma)^{-2}S_x]_x + \alpha(\beta S+\gamma)^{-2}S_x$ occurring in two-phase flow in porous media, *SIAM J. Appl. Math., 42*, 318-332,1982.

Forsyth, A.R., *Theory of Differential Equations,* Vol. 6, Cambridge Univ. Press, Cambridge, 1906.

Freeman, N.C., and J. Satsuma, Exact solution describing an interaction of pulses with compact support in a nonlinear diffusive system, *Phys. Lett. A, 138,* 110- 112, 1989.

Freeze, R.A., A stochastic-conceptual analysis of rainfall-runoff processes on a hillslope, *Water Resources Res., 16*(2), 391-408, 1980.

Fujita, H., The exact pattern of a concentration-dependent diffusion in a semi-infinite medium, 2, *Textile Res. J., 22,* 823-827, 1952.

Fulford, G. R., and P. Broadbridge, *Industrial Mathematics: Case Studies in Heat and Mass Transport,* Australian Mathematical Soc. Lecture Notes Series, Cambridge University Press, Cambridge, 2001.

Gardner, W.R. Some steady-state solutions of the unsaturated moisture flow equation with application to evaporation from a water table. *Soil Sci.,* 85, 228- 232, 1958.

Gautschi, W., Error Function and Fresnel Integrals, in *Handbook of Mathematical Functions,* (M. Abramowitz and I. E. Stegun, eds.), National Bureau of Standards, Washington DC, 1965.

Goodrich, D. C., J. M. Faurès, D. A. Woolhiser, L. J. Lane, and S. Sorooshian, Measurement and analysis of small-scale convective storm rainfall variability, *J. Hydrology 173*:283-308, 1995.

Green, W.A., and G.A. Ampt, Studies on soil physics: 1. The flow of air and water through soils, *J. Agricultural Science, 4,* 1-24, 1911.

Hastings Jr., C., *Approximations for Digital Computers,* Princeton University Press, Princeton NJ, 1955.

Haverkamp, R., J.-Y. Parlange, J.L. Starr, G.Schmitz, and C. Fuentes, Infiltration under ponded conditions: 3. A predictive equation based on physical parameters, *Soil Science, 149*(5), 292-300, 1990.

Haverkamp, R., P.J. Ross, K.R.J. Smettem, and J-Y. Parlange, Three-dimensional analysis of infiltration from the disc infiltrometer 2. Physically-based infiltration equation, *Water Resour. Res., 30,* 2931-2935, 1994.

Haverkamp, R., and M. Vauclin, A note on estimating finite difference interblock hydraulic conductivity values for transient unsaturated flow problems, *Water Resources Res., 15*(1), 181-187, 1979.

Hawkins, R.H., and T.W. Cundy, Steady-state analysis of infiltration and overland flow for spatially-varied hillslopes, *Water Resources Bull., 32*(2), 251-256, 1987.

Hillel, Daniel, *Fundamentals of Soil Physics,* Academic Press, New York, 1980.

Hofmann, L.L., W.R. Guertal, and A.L. Flint, Development and testing of techniques to obtain infiltration data for unconsolidated surficial materials, Yucca Mountain area, Nye County, Nevada, USGS Open-File Reprot 95-154, 23 pp., 2000.

Hopf, E., The partial differential equations $u_t + uu_x - \mu u_{xx}$, *Commun. Pure Appl. Math., 3,* 201-230, 1950.

Hussen, A.A., and A.W. Warrick, Alternative analyses of hydraulic data from the disc tension infiltrometer. *Water Resour. Res.,* 29, 4103-4108, 1993.

Jarvis, N.J., and Messing, I. Near-saturated hydraulic conductivity in soils of contrasting texture as measured by tension infiltrometers. *Soil Sci. Soc. Amer. J.,* 59, 27-34, 1995.

Jury, W.A., W.R. Gardner, W.H. Gardner, Soil Physics, fifth ed., J. Wiley, N.Y., 328 pp., 1991.

Kirchhoff, G., *Vorlesungen über der Theorie der Wärme,* Barth, Leipzig, 1894.

Knight, J. H., and J.R. Philip, Exact solutions in nonlinear diffusion, *J. Engineering Math., 8*, 219-227, 1974.

Li, R.M., M.A. Stevens, and D.B. Simons, Solutions to Green-Ampt infiltration equation, *J. Irrigation and Drainage Div., ASCE, 102*(2), 239-248, 1976.

Liu, M.-C., J.-Y. Parlange, M. Sivapalan, and W. Brutsaert, A note on the time compression approximation, *Water Resour. Res., 34*(12), 3683-3686, 1998.

Loch, R.J., Bourke, J.J., Glanville, S.F. and Zeller, L. Software and equipment for increased efficiency of field rainfall simulation and associated laboratory analysis. *Soil and Tillage Res.*, 45, 341-348. 1998.

Loch, R.J., Robotham, B.G., Zeller, L., Masterman, N., Orange, D.N., Bridge, B.J., Sheridan, G. and Bourke, J.J. A multi-purpose rainfall simulator for field infiltration and erosion studies. *Aust. J. Soil Res.*, 39, 599-610.

Logsdon, S.D., and Jaynes, D.B. Methodology for determining hydraulic conductivity with tension infiltrometers. *Soil Sci. Soc. Amer. J.*, 57, 1426- 1431, 1993.

Maller, R.A., and M.L Sharma, An analysis of areal infiltration considering spatial variability, *J. Hydrology*, 52, 25-37.

McWhorter, D.B., Infiltration affected by flow of air, *Hydrology Paper No. 49,* Colorado State University, Fort Collins, 43 pp., 1971

Mohanty, B.P., Ankeny, M.D., Horton, R., and Kanwar, R.S. Spatial analysis of hydraulic conductivity measured using disc infiltrometers. *Water Resour. Res.*, 30, 2489-2498.

Mohanty, B.P., R.S. Bowman, J.M.H. Hendrickx, and M.T. van Genuchten. New piecewise-continuous hydraulic functions for modelling preferential flow in an intermittent-flood-irrigated field. *Water Resour. Res.*, 33, 2049-2063, 1997.

Morel-Seytoux, H.J., Two-phase flows in porous media, in *Advances in Hydroscience, V. 9*, edited by V.T. Chow, pp. 119-202, Academic Press, New York, 1973.

Morin, J., Goldberg, D., and Segnier, I. A rainfall simulator with a rotating disc. *Trans. ASAE.*, 10, 74-77, 1967.

Mualem, Y., A new model for predicting the hydraulic conductivity of unsaturated porous media, *Water Resources Res., 12*, 513-522, 1976.

Mualem, Y., and S. Assouline, Modeling soil seal as a nonuniform layer, *Water Resources Res., 25(10),* 2101-2108, 1989.

Mutchler, C.K. and Hermsmeier, L.F. Review of rainfall simulators. *Trans. Am. Soc. Agric. Eng.*, 8, 67-69, 1969

Nielsen, D.R., J.W. Biggar, and K.T. Erh, Spatial variability of field measured soil-water properties, *Hilgardia, 42*, 215-259, 1973.

Parlange, J.-Y., W. Hogarth, P. Ross, M.B. Parlange, M. Sivapalan, G.C. Sander, and M.C. Liu, "A note on the error analysis of time compression approximation, *Water Resour. Res., 36*(8), 2401-2406, 2000.

Parlange, J.-Y., I. Lisle, R.D. Braddock, and R.E. Smith, The three-parameter infiltration equation. *Soil Science, 133*(6), 337-341, 1982.

Parlange, J.-Y., and R.E. Smith, Ponding time for variable rainfall rates. *Can. J. of Soil Science, 56*, 121-123, 1976.

Perroux, K.M., and White, I. Designs for disc permeameters. *Soil Sci. Soc. Amer. J.*, 52, 1205-1215, 1988.

Philip, J.R., Theory of infiltration 1. The infiltration equation and its solution. *Soil Science, 83*(5), 345-358, 1957a.

Philip, J.R., Theory of infiltration 2. The profile of infinity. *Soil Science, 83*(6), 435-448, 1957b.

Philip, J.R., The theory of infiltration 4. Sorptivity and algebraic infiltration equations. *Soil Science, 84*(3), 257-264, 1957c.

Philip, J.R. Absorption and infiltration in two- and three-dimensional systems, in *Water in the unsaturated zone. Vol. 1.* edited by R.E. Rijtema and H. Wassink, pp 503-525, UNESCO, Paris, 1966.

Philip, J.R. Steady infitration from buried point sources and spherical cavities. *Water Resour. Res.*, 4, 1039-1047, 1968.

Philip, J.R., Theory of infiltration, in *Advances in Hydroscience, Vol. 5*, edited by V.T. Chow, Academic Press, NY, pp. 215-296, 1969.

Philip, J.R., On solving the unsaturated flow equation: 1. The flux-concentration relation. *Soil Science, 116*(3), 328-335, 1973.

Philip, J.R., Reply to "Comment on steady infiltration from spherical cavities," *Soil Science Soc. of America J., 49*, 788-789, 1985.

Philip, J.R., The infiltration joining problem, *Water Resources Res. 23*(12), 2239- 2245, 1987.

Rawitz, E., Margolin, M., and Hillel, D. An improved variable-intensity sprinkling infiltrometer. *Soil Sci. Soc. Amer. Proc.*, 36, 533-535, 1972.

Reeves, M., and E.E. Miller, Estimating infiltration for erratic rainfall, *Water Resour. Res., 11*(1), 102-110, 1975.

Reisenauer, A.E., Methods for solving problems of multidimensional partially saturated steady flow in soils, *J. of Geophysical Research, 68*, 5725-5733, 1963.

Reynolds, W.D., and Elrick, D.E. Ponded infiltration from a single ring 1. Analysis of steady flow. *Soil Sci. Soc. Amer. J.*, 54, 1233-1241, 1990.

Richards, L.A., Capillary conduction of liquids through porous mediums. *Physics, 1*, 318-333, 1931.

Rogers, C., M.P. Stallybrass, and D.L. Clements, On two phase filtration under gravity and with boundary infiltration: Application of a Bäcklund transformation, *Nonlinear Anal. Theory Methods Appl., 7*, 785-799, 1983.

Rosen, G. Method of the exact solution of a nonlinear diffusion-convections equation, *Phys. Rev. Lett., 49*, 1844-1846, 1982.

Ross, P.J. and Bridge, B.J. A portable, microcomputer-controlled drip infiltrometer. I. Design and operation. *Aust. J. Soil Res.*, 23, 383-391. 1985.

Rubin, J. Theory of rainfall uptake by soils initially drier than their field capacity and its applications. *Water Resour. Res.* 2, 739-749.

Sander, G.C., J-Y. Parlange, V. Kühnel, W.L. Hogarth, D. Lockington, and J.P.J. O'Kane, Exact nonlinear solutions for constant flux infiltration. *J. of Hydrology, 97*, 341-346, 1988.

Scotter, D.R., Clothier, B.E., and Harper, E.R. Measuring saturated hydraulic conductivity and sorptivity using twin rings. *Aust. J. Soil Res.*, 20, 295-304, 1982.

Sharma, M.L., G.A. Gander, and C.G. Hunt, Spatial variability of infiltration in a watershed, *J. of Hydrology 45*, 101-122, 1980.

Singh, V.P., *Kinematic wave modeling in water resources: Surface water hydrology*, John Wiley and Sons, Inc., New York, 1996.

Šimnek, J., T. Vogel, and M. Th. Van Genuchten, The SWMS_2D code for simulating water flow and solute transport in two-dimensional variably saturated media, Version 1.2, Research Report No. 132, U.S. Salinity Laboratory, USDA, ARS, Riverside, CA, 1994.

Sivapalan, M., and P.C.D. Milly, On the relationship between the time condensation approximation and the flux-concentration relation, *J. of Hydrology 105*, 357-367, 1989.

Sivapalan, M., and E.F. Wood, Spatial heterogeneity and scale in the infiltration response of catchments, Ch. 5 in *Scale Problems in Hydrology,* edited by V.K. Gupta, I. Rodriguez-Iturbe, and E.F. Wood, Reidel, Hingham, MA, 81-106, 1986.

Smettem, K.R.J., and Collis-George, N. Prediction of steady-state ponded infiltration distributions in a soil with vertical macropores. *J. Hydrol.*, 79, 115-122, 1985.

Smettem, K.R.J. Characterization of water entry into a soil with a contrasting textural class: spatial variability of infiltration parameters and influence of macroporosity. *Soil Sci.*, 144, 167-174, 1987.

Smettem, K.R.J., and Clothier, B.E. Measuring unsaturated sorptivity and hydraulic conductivity using multiple disc permeameters. *J. Soil Sci.*, 40, 563- 568, 1987.

Smettem, K.R.J. and Ross, P.J. Measurement and prediction of water movement in a field soil: the matrix-macropore dichotomy. *Hydrol. Processes,* 6, 1-10, 1992.

Smettem, K.R.J., Parlange, J-Y, Ross, P.J., and Haverkamp, R. Three-dimensional analysis of infiltration from the disc infiltrometer: 1. A capillary-based theory. *Water Resour. Res.,* 30, 2925-2929, 1994.

Smettem, K.R.J., Ross, P.J., Haverkamp, R., and Parlange, J-Y. Three-dimensional analysis of infiltration from the disk infiltrometer 3. Parameter estimation using a double-disk tension infiltrometer. *Water Resour. Res.*, 31, 2491-2495, 1995.

Smith, R.E., Approximate soil water movement by kinematic characteristics, *Soil Science Society of America Journal, 47*(1), 3-8, 1983.

Smith, R.E., Analysis of infiltration through a two-layer soil profile, *Soil Science Society of America J., 54*(5), 1219-1227, 1990.

Smith, R.E., Technical note: Rapid measurement of soil sorptivity, *Soil Science Society of America J., 63*(1), 55-57, 1999.

Smith, R.E., and J.-Y. Parlange, A parameter-efficient hydrologic infiltration model, *Water Resources Res., 14*(3), 533-538, 1978.

Smith, R.E., C. Corradini, and F. Melone, Modeling infiltration for multistorm runoff events. *Water Resources Res., 29*(1), 133-144, 1993.

Smith, R.E., and R.H.B. Hebbert, A Monte-Carlo analysis of the hydrologic effects of spatial variability of infiltration, *Water Resour. Res., 15*(2), 419- 429, 1979.

Smith, R.E., D.C. Goodrich, and D.A. Woolhiser, Areal effective infiltration dynamics for runoff on small catchments, in *Proc., Trans. 14th Int. Congr. of Soil Sci.,* Vol. I, International Society of Soil Science, Kyoto, Japan, pp. I-22 - I-27, 1990.

Smith, R.E., and D.A. Woolhiser, Overland flow on an infiltrating surface, *Water Resources Res., 7*(4), 899-913, 1971.

Sposito, G., Recent advances associated with soil water in the unsaturated zone, *Reviews of Geophysics: Supplement, "U.S. national report to IUGG,"* pp.1059-1065,1995.

Stone, J.J., R.H. Hawkins, and E.D. Shirley, Approximate form of Green-Ampt infiltration equation, *J. of Irrigation and Drainage Engineering, 120*(1),128- 137, 1994.

Storm, M. L., Heat conduction in simple metals, *J. Appl. Phys.*, 22, 940-951, 1951.

Svinolupov, S. I., Second-order evolution equations with symmetries, *Usp. Mat. Nauk.*, 40, 263-4, 1985.

Turner, N.C., and Parlange, J-Y. Lateral movement at the periphery of a one-dimensional flow of water. *Soil Sci.,* 118, 70-77, 1974.

Talsma, T. and J.-Y. Parlange, One-dimensional vertical infiltration, *Aust. J. Of Soil Res., 10*(2), 143-150, 1972.

Vandavaere, J-P., Peugeot, C., Vauclin, M., Angulo-Jaramillo, R., and Lebel, T. Estimating hydraulic conductivity of crusted soils using disc infiltrometers and minitensiometers. *J. Hydrol.,* 188, 203-223, 1997.

Vandavaere, J-P., Vauclin, M., and Elrick, D.E. Transient flow from tension infiltrometers: 1. The two-parameter equation. *Soil Sci. Soc. Amer. J.*, 64, 1263-1272, 2000a.

Vandavaere, J-P., Vauclin, M., and Elrick, D.E. Transient flow from tension infiltrometers: II. Four methods to determine sorptivity and conductivity. *Soil Sci. Soc. Amer. J.* 64, 1272-1284, 2000b.

van Genuchten, M. Th., A closed-form equation for predicting the hydraulic conductivityof unsaturated soils, *Soil Science Soc. of America J., 44*(5), 892- 898, 1980.

Viera, S.R., D.R. Nielsen, and J.W. Biggar, Spatial variability of field-measured infiltration rate, *Soil Science Soc. Of America J., 45*(6), 1040-1048, 1981.

Warrick, A.W. Models for disc infiltrometers. *Water Resour. Res.,* 28, 1319- 1327, 1992.

Warrick, A.W., and P. Broadbridge, Sorptivity and macroscopic capillary length relationships, *Water Resour. Res., 28*, 427-431, 1992.

Weir, G.J. Steady infiltration from small shallow circular ponds. *Water Resour. Res.,* 23, 733-736, 1987.

White, I. Tillage practices and soil hydraulic properties: why quantify the obvious? in *National Soils Conference Review Papers*. Edited by J. Loveday, pp 87- 126., Aust. Soc. Soil Sci. Canberra, Australia, 1988.

White, I., and P. Broadbridge, 1988, Constant rate rainfall infiltration: A versatile nonlinear model 2. Applications of solutions, *Water Resour.Res., 24*(1), 155- 162, 1988.

White, I., D.E. Smiles, and K.M. Perroux, Absorption of water by soil: The constant flux boundary condition, *Soil Science Society of America J., 43*(4), 659-664, 1979.

White, I, and M.J. Sully, Macroscopic and microscopic capillary length and time scales from field infiltration. *Water Resour. Res. 23*(8), 1514-1522, 1987.

White, I., Sully, M.J. and Melville, M.D. Use and hydrological robustness of time to incipient ponding. *Soil Sci. Soc. Amer. J.* 53, 1343-1346, 1989.

White, I., Sully, M.J., and Perroux, K.M. Measurement of surface-soil hydraulic properties: Disk permeameters, tension infiltrometers and other techniques. In *Advances in Measurement of Soil physical Properties: Bringing Theory into Practice*., edited by G.C. Topp et al., pp 69-103, SSSA Special Publication no. 30. Madison, WI, USA, 1992.

Wooding, R.A. Steady infiltration from a circular pond. *Water Resour. Res.*, 4, 1259-1273, 1968.

Woolhiser, D.A., Search for physically based runoff model - A hydrologic El Dorado?" *J. of Hydraulic Engineering, Amer. Soc. of Civil Engineers, 122*(3), 122-129, 1996. [see also Closure in *123*(9), 829-832, 1997]

Woolhiser, D.A., and D.C. Goodrich, Effect of storm rainfall intensity patterns on surface runoff, *J. Hydrology, 102*, 335-354, 1988.

Woolhiser, D.A., R.E. Smith, and J-V. Giraldez, Effects of spatial variability of saturated hydraulic conductivity on Hortonian overland flow, *Water Resour. Res., 32*(3), 671-678, 1996.

Wu,L., L. Pan, M.J. Robertson, and P.J. Shouse, Numerical evaluation of ring-infiltrometers under various soil conditions, *Soil Science, 162*(11), 771-777, 1997.

Youngs, E.G., An estimation of soprtivity for infiltration studies from moisture moment considerations, *Soil Science, 106*(3), 157-163, 1968.

Zaidel, J., and D. Russo, Estimation of finite difference interblock conductivities for simulation of infiltration into initially dry soil, *Water Resources Res., 28*(9), 2285-2295, 1992.

Zeglin, S.J., and I. White, Design for a field sprinkler infiltrometer. *Soil Science Society of America J., 56*, 1129-1133, 1992.

Zhang, R. Infiltration models for the disk infiltrometer. *Soil Sci. Soc. Amer. J.,* 61, 1597-1603, 1997.

Index